BEES & HONEY
FROM FLOWER TO JAR

BEES & HONEY
FROM FLOWER TO JAR

Michael Weiler

Floris Books

Translated by David Heaf

First published in German as *Der Mensch und die Bienen —
Betrachtungen zu den Lebensäußerungen des Bien*
by Verlag Lebendige Erde, Darmstadt
First published in English in 2006
Third printing 2011

© 2003 Verlag Lebendige Erde, Darmstadt
This translation © Floris Books, 2006

All rights reserved. No part of this publication may
be reproduced without the prior permission of
Floris Books, 15 Harrison Gardens, Edinburgh
www.florisbooks.co.uk

British Library CIP Data available

ISBN 978-086315-575-8

Printed in Great Britain
by Page Bros (Norwich), Ltd

Contents

Foreword   11

I. OBSERVATIONS   13
  1. Meeting   13
     Flower visit 13
     Pursuit 16
  2. Flight Observation   17
     The orientation flight of young bees 19
     The flight of the drones 19
     A hive swarms 21
     Queen flight 23
     Drone congregation areas 24
  3. The Swarm   26
     The swarm cluster 26
     Catching a swarm 29
     Hiving the bees in a new home 31

II. 'INSIDES'   35
  4. Building Comb   35
     Shaping space 37
     Beehives 38
  5. Bee Development   40
     Egg laying 41
     Larvae 42
     Pupation 44
     Drone brood 47
     Bees hatching 48
  6. Bee Life   51
     House bees — flying bees 51
     Development from the inside ... 53
     ... to the outside 54
     Metamorphosis of the glands 55
     Interpenetration 57
  7. Summer Impressions   58
     Bee dances 59

III. DEVELOPMENTS 65
   8. Queen Cells: Queen Development 66
   9. Maturity 71
      Reduction — departure of the drones 72
      Wintering 74
  10. Winter Inspections 78
      Reading the debris 79
      An emergency 80
  11. The Varroa Mite: a Parasite of Bee Colonies 81
  12. Midwinter 85
  13. Bee Christmas 86

IV. BUILD-UP 87
  14. A Time of Awakening 87
      Cleansing flight 87
  15. Getting through Spring 89
      Dangers 89
  16. Winter Bees: Summer Bees 91
      Help from the beekeeper 92
  17. Preparing for the Honey Harvest 94
  18. Summer Inspections 97
      Swarm control 98
      Swarm mood 100
      Swarm control in beekeeping 102

V. A CHALLENGE 105

VI. THE GIFT FROM THE BEES 109
  19. Honey 110
      People, honey and bees 110
      Honey processing by bees 112
  20. What Is the Significance of Honey
      for Bees and People 116
      Summer honey 116
      'Winter honey' 117
      The importance of honey to people 118

CONTENTS

The value of honey 119
What should honey really cost? 120
21. Royal Jelly: Liquid Feed for Queen Bees 121
    Pollen and bee bread 121
    What is bee bread used for in colonies? 123
    Royal jelly 124

Postscript 127
Acknowledgments 129
Selected Bibliography 131
Appendix: Demeter Beekeeping 133
Contact Details 137

Epirrhema

Müsset im Naturbetrachten
Immer eins wie alles achten;
Nichts ist drinnen, nichts ist draußen:
Denn was innen, das ist außen.
So ergreifet ohne Säumnis
Heilig öffentlich Geheimnis.

Freuet euch des wahren Scheins.
Euch des ernsten Spieles:
Kein Lebendiges ist ein Eins,
Immer ists ein Vieles.

Johann Wolfgang von Goethe

In contemplating Nature's being,
Know the One as many, seeing
In and outer coinciding,
Nothing in from out dividing.
Open secret, revelation!
Grasp it without hesitation.

Free of seeming truth's confusion,
Revel in the serious game!
Separateness is the illusion —
One and many are the same.

*English rendering by Aldyth Morris**

* from Goethe's Botanical Writings, Bertha Mueller, University of Hawaii Press, 1954

Im Regen geschrieben

Wer wie die Biene wäre,
die die Sonne
auch durch den Wolkenhimmel fühlt,
die den Weg zur Blüte findet
und nie die Richtung verliert,
dem lägen die Felder in ewigem Glanz;
wie kurz er auch lebte,
er würde selten
weinen.

*Hilde Domin, 'Nur eine Rose als Stütze,' 1997,
Fischer Tb*

Written in the rain

Whoever would become like a bee
who feels the sun
even through a cloudy sky,
who finds her way to a flower
and never gets lost,
to him the fields would lie in eternal radiance;
however short his life
he would rarely ever
complain.

Foreword

It is hard to find a description of what moves us when we meet bees that is as vivid and has such empathy as that in Hilde Domin's poem. People have always been fascinated by bees although, or perhaps because, the life of these insects takes place largely out of our sight.

There are certainly only a few points of direct contact between people and bees. We may see them in spring and summer when they visit fruit blossom or meadow flowers. Our attention on those occasions is perhaps first caught by the constant humming. Only when we look closer do we discover the bees busy inside the flowers.

Someone out for a walk might also come upon an apiary or, more rarely, a bee house. The more venturesome who go closer to the hive

Bee gathering nectar on snowberry *(Symphoricarpos)*
[From Zander and Weiss]

entrances will be able to see the purposeful comings and goings of the bees and the colourful pollen baskets (corbiculae) on their legs.

And we meet the bees again with the results of their work, albeit indirectly through our sense of taste, when we eat honey. All honeys taste different from one another depending on the nectar or honeydew source that the bees fed on when they made it.

But, for most people, what happens between the flower visit and enjoying honey remains hidden from view. Only the beekeeper in charge of the bees has the special privilege of penetrating the microcosm of a bee colony. He can directly observe the life processes and all the behaviour of the individual bees. Even professional beekeepers of long experience are time and again deeply moved and filled with childish wonder when they open a beehive.

Bees and Honey is more than a book on beekeeping. Michael Weiler addresses himself to anyone who is at all intrigued by bees and would like to learn more about these little insects.

The author is a beekeeper and teacher, an experienced observer and faithful narrator. He takes the reader with him to his bees and reports on everything that can be seen and experienced there. This allows readers to participate in the mysterious and wonderful life of the bees.

In the end, not only is our thirst for knowledge quenched but also our outlook is imperceptibly and subtly changed. Readers will explore the bee world with new eyes and new feelings.

Günter Friedmann (Master Beekeeper)

I. OBSERVATIONS

1. Meeting

On a mild, dry, sunny day in spring a walker is standing under a flowering fruit tree enjoying, perhaps with closed eyes, the scent that is wafting from it.

A loud buzzing awakens his attention and he notices a fat, flying insect busying itself, hurrying in wide curves from flower to flower. After a while it flies away again quickly. At this time of year it must have been a queen bumblebee.

Flower visit

But as the observer can still hear a gentle humming he searches for its source. On looking closer he sees that a lot of smaller insects are active in the tree in the same way. All the while some are leaving and others are arriving, usually from the same direction, only to busy themselves in the flowers.

More detailed observation reveals that, what he now recognizes as bees, are trying to get something from the flowers. A longish,

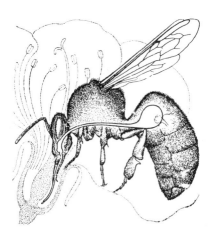

Worker bee showing the honey crop or honey stomach [From Büdel and Herold]

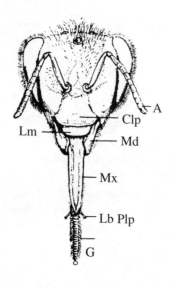

Head and mouth parts of worker bee:
A = antenna
Clp = clypeus
Lm = upper lip (labrum)
Md = upper mandible
Mx = lower mandible (maxilla)
Lb Plp = labial palp (tongue feeler)
G = tongue (glossa)

Tip of tongue (glossa) showing 'spoon' (flabellum)

[Both figs. from Büdel and Herold]

narrow, mobile organ emerges from between two small, pincer-like jaws at the front of their heads and plunges deep inside a flower. On enlargement this would show a complex proboscis-like tongue that ends in the shape of a little spoon which the bee uses to take up the nectar provided by the flower.

At the same time other bees are repeatedly dusting themselves with flower pollen. It stands out brightly against the short hair on their furry bodies. While a bee is flying to the next flower, it quickly sweeps its three pairs of legs alternately over its body, often hovering briefly in the air. This happens so fast that it is hardly noticeable. As a result a small lump forms on each rear leg. Enlargement shows that on each rear leg there are devices like combs and rakes, and a fold in which the lumps of pollen are kept — 'pollen baskets' as beekeepers call them.

1. MEETING

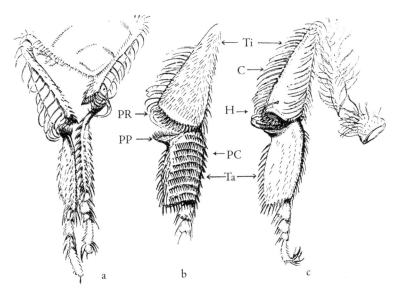

Pollen collection apparatus on the rear legs of worker bees
[from Dorothy Hodges, 1952]

a = rear view
b = left leg viewed from inside
c = right leg viewed from outside
Ti = tibiae
Ta = tarsi
C = pollen basket (corbicula)

H = single hair
PR = pollen rake
PP = pollen press
PC = pollen comb

[From Zander and Weiss]

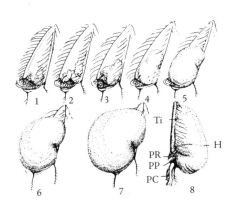

Forming the pollen load
[from Hodges]

Bee scraping pollen into loads

Pursuit

The observer, curiosity aroused, now tries to follow the flight path of the bees leaving the tree.

On a foraging flight a honey bee flies at a height of between 2 and 8 metres above the ground and, depending on terrain and wind conditions, at a speed of about 6–8 m/sec (14–18 mph). It carries 0.05 to 0.07 g of nectar, or about half of its body weight, in its honey crop or honey sac or honey stomach (hereafter referred to simply as 'crop'), which is a special compartment of its anterior alimentary tract.

When bees are constantly flying to the same foraging site such as a fruit tree in blossom, they create a sort of 'flight path' in the air. They make their way along this in a gently oscillating flight to their goal, never in a dead straight line. The unladen bees fly somewhat higher and faster, and those returning with nectar and pollen somewhat lower and more slowly.

With a lot of patience, keen eyes and a bit of luck, it is possible to discover where they are heading to or coming from. Usually this will be an apiary or a bee house, perhaps on a woodland margin or in a small sheltered clearing or in somebody's garden. As we approach such places the number of bees in flight increases. The humming is noticeable again, now louder. It can be shown that it is at a constant pitch which has been measured as 235 Hertz (Griziwa, J., Büdel and Herold, p. 171)

Bees coming in from all directions are impatiently heading for the hive entrances where, for the time being, they disappear from the observer's sight. Others emerge in their place and spiral up into the air, or set off immediately to an unknown destination with a gently oscillating flight. This constant flying back and forth has a busy purposefulness about it. It is sometimes sporadic, though nevertheless rhythmic.

2. Flight Observation

Nearby a partly dammed-up stream catches our attention. Water is glistening on the mossy roots of the bushes lining its bank, and bees are constantly alighting on it. Their abdomens move rhythmically. As if pumping they contract and expand a little in rapid succession. After a short time they fly back to the apiary. They are obviously collecting water from the brook.

Not far away is a chestnut tree on whose drooping branches there are still some fat buds with a resinous gleam. The bees are busy at work there too. They are biting off pieces of the bud resin with their mandibles and transferring them with rapid, skilful movements as little droplets on their rear legs, to the place where the pollen gatherers store their pollen loads. This bud resin is rather sticky and it is amazing how the bees manage to avoid getting sticky themselves.

The observer can now take stock of his observations with the following summary:

— the meeting at the flowering tree;
— the deliberate flight movements, leaving a central point and leading back to it again;

Bees in flight [From Doering and Hornsmann]

Bee drinking [From Doering and Hornsmann]

— the repetitious activity of gathering various things in many visits to the surroundings of the apiary;
— discovering (with the help of further investigation or appropriate reading) that the insects under observation are physically and organically adapted especially for their activity.

By summarizing this temporally, and thus enhancing our idea of what has been observed of these processes, we obtain a mobile picture: a movement from a centre to an invisible periphery and back again. The movement is constantly oscillating, almost breathing, interpenetrating itself, reaching out and bringing back in again. The form of the sphere embraced is in constant change, larger, smaller, now tending more one way, now switching to the other, depending on how attractive the goal is that draws it the force living at its centre.

In this way we get an initial impression of something limb-like that is reaching out over and over again beyond the hive in order to accomplish something out there and then fetch it back to itself.

2. FLIGHT OBSERVATION

The orientation flight of young bees

With repeated visits to the apiary observers will be able to extend their experience of different kinds of bee flight. From Easter onwards on dry, warm mornings around eleven o'clock, and again in the afternoons at around four, large numbers of bees can be seen in front of the hives behaving totally differently from those engaged so purposefully in gathering.

Lots of bees come out of the entrance, slowly rise into the air in front of the hive and, while flying, turn their heads towards the front wall of the hive immediately above the entrance. Then they fly up and down, back and forth, landing repeatedly on the landing board or the front of the hive, and taking off again soon after. The radii of their flights get bigger and the bees fly round the immediate vicinity of the apiary in broad oscillating movements, like lemniscates or figures-of-eight. They soon return and disappear into the hive entrance.

Beekeepers call this phenomenon the orientation flight, or occasionally the play or practice flight. Young bees, still completing their development as hive or house bees, fly like this in the environs of their hive. The constant to and fro of the foraging flights continues unaffected.

The flight of the drones

However, the visitor notices yet another surprising thing. With a loud buzz, deeper in pitch than the general hum of the apiary, a fat insect approaches fast and disappears into a hive entrance. From other hives, especially those with heavy flight traffic, similar droners emerge, move heavily across the alighting board and take off with a loud drone. At first they rise somewhat clumsily, almost staggering in the air, but then they ascend with a powerfully direct flight and disappear into the environs. Despite their size and their almost droning flight sound, it is difficult to keep them in sight for long, because they fly away so quickly. These fliers are called 'drones,' because of the noise they make.

Closer observation reveals that even externally they are extremely different from the foragers. First we notice that drones are both longer and fatter. Their compound eyes are much bigger and almost cover the whole of their heads. Detailed investigation has shown that the mobile antennae that grow from the 'middle' of the 'face'

Drone [From Doering and Hornsmann]

have one segment more than those of foragers; eleven instead of ten. Also the mouth parts of the drones are less developed than those of the workers. The proboscis is significantly shorter and less differentiated; the mandibles of the mouth parts are smaller and shorter. The furry thorax of drones is fatter and the pairs of wings are so big that they almost extend beyond the end of the abdomen. The three pairs of legs seem barely capable of supporting the body and they are less differentiated in their outer segments. They also lack the forager's pollen gathering apparatus (rakes, combs etc). The abdomen is longer too with the effect that drones are about half as long again as workers. The observer can pick up a drone without risk of being stung and even stroke it. They posses neither sting nor poison gland. Instead, with suitable pressure on the abdomen, a milky-white strangely-shaped organ emerges that biologists call the male sex-organ.

Perhaps the observer is now struck by the curious fact that they have never before seen a drone in the open even though here in the apiary there are many of them in flight. We see them neither on flowers nor inside our house windows, where sometimes a worker bee out hunting is trapped after taking a wrong turn. And yet it is a fact that they leave the apiary and come back again. Closer investigation would reveal that the drones go 'roaming' within a certain radius and are always admitted to any colony and have their needs attended to by it, even at another apiary. But where do they go and where do they stay during their flights that last on average from forty-five minutes to an hour? We shall come back to this question later.

A hive swarms

Now the inquiring visitor is spellbound by another event. It is close to noon. The sun is almost at its zenith. For some time there has been a certain amount of disturbance at one of the hives. Lots of drones have been flying in and out. The orientation flight has just finished, but quite a lot of bees are scurrying around the hive entrance, going in and out or rocking backwards and forwards as if scrubbing ('washboard movement'). Suddenly the busy foraging flight is replaced by a new activity. Masses of bees are pouring out of the hive entrance almost like a stream of water, tripping over and bumping into one another and immediately rising into the air, which is soon filled with a loud buzzing. While the outflow has still not finished, the bees that have taken off fly high into the sky. They do not head purposefully to some point like the foragers, but instead fly round in large lemniscate loops near to, or further away from, the apiary; some quickly, like streaks lightning in the sunshine, others slowly and peacefully. It seem as if there is no spot in the whole surroundings that is not visited or flown through by a bee. They alight on the grass, bushes and trees; crawl around briefly and take off again. A lot of bees even approach the astonished, almost shocked, onlookers and crawl around on them as if looking for something — a little dreamily almost, even going into their noses and ears, tickling peaceably — and then slowly going away again.

Meanwhile the flow from the hive entrance has dried up. At the moment there is almost no more activity there. Instead the observer notices a small collection of bees busy on a tree a few metres from the hive. On cautiously approaching — there are still bees crawling around on the grass — he notices that more and more bees are landing there. He may even notice that a bee with gleaming red legs, which seems longer and more slender than the rest, is being covered by the new arrivals. By now a cluster of bees about the size of a pear is hanging on the branch. On the outside of it are quite a number of

Bees fanning with their scent glands exposed [Photo: Tim Kraus, Kassel, 1996]

bees with their abdomens pointing upwards into the air and the tip bent downwards somewhat. This is exposing a light-brown, slightly everted area between the penultimate tergite segments. They are fanning their wings powerfully without flying away and thus producing an easily noticeable flow of air. The bee literature, or a beekeeper, will tell us that the light-brown area is the Nasonov scent gland and that the bees with their bottoms in the air are 'scenting' by blowing a current of scent out into the air. This attracts the bees still flying around to the cluster.

It is now at least five minutes since swarming began. At the hive entrance the foraging traffic is almost back to normal and barely distinguishable from that at other hive entrances. The buzzing hum of the swarming bees is creating a uniform sound picture. The number of bees flying to the swarm cluster and landing on it is increasing and it is slowly getting bigger. More bees join in the scenting, and the flight paths of those still in the air are shortening. About fifteen minutes from the start almost all the bees are in the swarm cluster which has now grown to about the size of a rugby ball. A swarm of this size weighs about 2.5 kg and contains about 18,000 to 20,000 bees and the stores of honey they consumed before swarming. Closer

2. FLIGHT OBSERVATION

Swarm cluster [Photo: author]

observation shows that there are also drones in the swarm and bees with pollen baskets which have obviously joined the swarm immediately after returning from their foraging flight. After twenty minutes, peace returns once more to the apiary; bee traffic is as before. Seen from a short distance away the swarm is hanging apparently motionless in the tree. The casual passer-by would probably not notice it.

What a swarm does next is discussed in Chapter 3.

Queen flight

When our observer returns to the apiary ten to twelve days later he first looks at the hive that threw the swarm. (Beekeepers say that a colony from which a swarm has issued has 'thrown' a swarm). But everything about it looks the same as the other hives. The orientation flight of the young bees has just finished; drone flights have increased. After a while, when the sun has just passed its zenith, a bee appears at the hive entrance that looks different from the by now familiar workers and drones. She seems slender and is longer than the others. Her abdomen extends well beyond her wings, and her legs look somewhat longer and have a reddish gleam. She immediately takes off. From her distinctive size it is easy to distinguish this bee in flight from the others and she can quite easily be kept in view

Queen bee [From Zander and Weiss]

for a time. She flies in a few arcs in front of the hive, climbs higher and disappears quickly in a specific direction. The observer searches carefully to see if he can see any other bees of this kind, but without success.

Only after about half an hour has passed does he notice this large bee reappear. She lands again on the alighting board; the other bees on it come up to her and touch her with their feelers and forelegs and stroke her with their proboscises. A white appendage is hanging out of the tip of her abdomen, gleaming somewhat moistly like a little pennant. She walks in through the entrance and disappears.

In talking to the beekeeper, or through reading, the enquirer discovers that he has seen a queen bee, a virgin queen, making a mating flight.

Drone congregation areas

So far nobody has been able to observe the 'open air' mating flight of a queen bee. But conditions that give an idea of this event can be arranged experimentally.

For a long time it was not known why there were drones in a bee colony and people did not know where they flew to. Reports of so-called 'drone congregation areas,' supposedly somewhere in the vicinity, were always cropping up without anyone knowing exactly what they were. More recently it has become possible to ascertain certain facts about these places.

2. FLIGHT OBSERVATION

Apparently every year from the end of April to the end of August the flying drones of the colonies in a neighbourhood congregate — always in the same places. There, in dry, warm weather they fly around within a specific area between about eleven o'clock in the morning and five o'clock in the afternoon. In fact, they fly back and forth so fast that they cannot be spotted. These places can be recognized only by the low hum of the drones, but they can be detected briefly if a stone is thrown up into the air because a few drones will fly towards it. If a lure is made with a caged queen fixed to a helium balloon attached to a string, the height and extent of such areas can be determined. As soon as she is pulled out of the congregation area, no more drones will follow. In such an artificially contrived way it has been possible to film the mating of a queen.

From these observations it can be concluded that in her mating flight, often several in succession, the queen flies to a drone congregation area and is there chased by the mature drones and mated. The drone that reaches the queen clasps her abdomen from above with his hind legs. By squeezing all the air and fluids in his body into his abdomen he everts his genitalia. During the ensuing copulation this is firmly locked into the opening of the queen's genitalia at the tip of her abdomen and this injects the seminal fluid into the queen's spermatheca. The drone tips over backwards and falls away; his genitalia are torn out and remain hanging from the genital orifice of the queen. The drone falls dead to the ground.

Mating can happen with several drones in succession, on average between five and fifteen. Each subsequent drone removes the 'mating sign' of its predecessor. The queen brings the last sign back to the hive.

3. The Swarm

Twenty minutes after the swarm has left the hive, calm has returned to the apiary. The mood of excitement that filled the whole area through the thousands upon thousands of swarming bees has once again returned to a constant hum of departing and returning foragers, modified only by the droning buzz of the drones. The bees have now gathered at a single point; after the out-swarming, the concentration — diastole and systole on a grand scale.

The swarm cluster

The swarm is hanging motionless on a branch in the shadow of the crown of the tree and is not immediately obvious. Certainly a casual passerby would be unlikely to notice it. Even a beekeeper who was not there to observe the hive swarming would have to spend some time looking, if the cluster was not hanging at a place where swarms had frequently landed in the past. As a matter of routine at swarming time, beekeepers scan the trees and bushes round the apiary as soon as they arrive, but they can also discover swarms because they notice a certain 'irregularity' which is introduced into the usual 'order' of the surrounding vegetation. At the moment of 'contact' perhaps an inner perception happens, something like an inner stirring, like an encounter.

But beekeepers can also have a clear overall view of a swarm, and for the time being we shall take a look at one of these.

After a short time, the observer notices on the outside of the swarm cluster a few bees who are behaving in a particular way. A bee walks over the cluster for a short distance in a straight line while at the same time rapidly waggling her abdomen from side to side. Then she returns in a semicircle without waggling her tail. Back at the starting point she repeats the waggle along the short straight line and this time returns to the starting point in a semicircle in the opposite direction to the first. She repeats this for several minutes each time with the semicircular path alternating right and left of the straight line (making a figure of eight, see also page 61). After watching this for some time it becomes clear that the straight line always points in the same direction and the walk maintains a precise rhythm, for example, five circuits in fifteen seconds.

3. THE SWARM

Swarm cluster [Photo: author]

Longer study reveals that other bees show the same behaviour only with different rhythms and directions of the 'wagtail' walk. Whereas the waggle dance of one bee always points upwards, another always points almost horizontally and the frequency of its circuits is lower. A third bee is observed pointing downwards to the left and making more frequent circuits than the others. Her waggle seems more vigorous, intense. Other bees go up to this one and

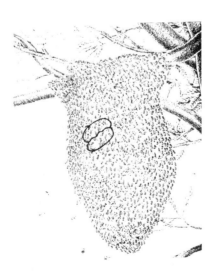

Swarm of bees, illustrating the pattern of the waggle dance [From Ruttner]

touch her with their antennae and briefly follow her after which they disappear into the cluster or fly away. (More will be said about this bee behaviour, also called the 'bee dance,' in Chapter 7, where we consider the bee dance as an inner expression of the relationship of a bee colony to its surroundings.)

Busy flight traffic gradually develops from the suspended swarm as several bees take off from its surface, whereas others use a hole that opens in it at the bottom end forming a kind of hive entrance. The cluster contracts and is less active at lower temperatures.

If it rains, the bees on the outside arrange themselves in layers like roof tiles with their heads facing upwards and their wings and abdomens downwards and outwards. It can rain for a long time without water penetrating the inside of the cluster.

If we watch the dance behaviour over a longer period, perhaps for several days, we will notice other features. Gradually an increasing number of dancing bees on different parts of the swarm-cluster point in the same direction, for example, in the third direction described above, but at the same time several dance in the first direction too. The second has disappeared completely. Other directions of movement are discovered and observed repeatedly but they eventually disappear.

One particular direction of the waggle dance begins to predominate, possibly after two days, and is then performed in the same way by the majority of the dancing bees. Suddenly the behaviour of the swarm changes. Vibrating their wings, the bees run around very fast seemingly in a muddle. Things start to loosen up, the cluster starts to disperse. In a short time the bees leave the place they started from and fly away, all in the same direction, in an elongated cloud. If they do not fly too far, the observer may be able to follow them and can watch how they gather at a new place and there disappear, perhaps through a hole or a crack, into a cavity. This could be a hollow tree or a crack in a rock; but it could also be the false ceiling of a half-timbered house or a cavity wall. Other possibilities include large empty nest-boxes, chimneys and, not least, an unoccupied hive left open at another apiary.

At the opening of this cavity a pattern of flight activity develops that is similar to that already described for the hive in the apiary. What is going on in the cavity is out of sight and for the time being can only be imagined by the inquiring observer.

3. THE SWARM

Catching a swarm

Only if a beekeeper 'takes' a swarm is it possible to follow its subsequent development. For this he needs first of all an empty container, which could be a cardboard box, a tub or anything similar that is bee-tight, closable, and allows sufficient ventilation. It is also useful to have a feather duster (for example, a goose wing) or a bee brush. It also helps to spray the outside of the cluster with a little water as this causes it to contract.

The beekeeper holds the box with one hand under the quietly suspended bees that seem to be snuggled up to one another. Then he gives the branch a vigorous shake downwards, whereupon the cluster drops into the box like a ripe fruit and there it flows around like thick soup. The few bees still hanging onto the branch are brushed, with a few short strokes, in with the rest. The box is then closed.

Once it is certain that almost all the bees are in the box then it can be taken away. But if a lot of bees took off or missed the box when the swarm was knocked down, the beekeeper places it on the ground

Flight entrance of a wild colony in a tree with a curtain of propolis in front (Entrance diameter: 6 cm) [From Zander and Weiss]

Swarm catching box with the swarm that has been knocked into it.
[Photo: author]

nearby, preferably in a shady place. He then opens a flight entrance for the bees and perhaps ensures that any breeze there may be is blowing from the box to the place where the swarm was hanging. For now the bees in the box are starting to roar and fan scent again. Those still outside are gathering at the starting point on the branch or flying round in the familiar lemniscatory loops. While they do this they are able to detect the scent and soon find their way to the swarm in the box.

That evening, at the latest, the swarm that has settled in the box can be taken away. The beekeeper places it in a cool, dark and quiet place, for instance, in a cellar. In this situation the bees will hang from the top of the box thus re-forming the swarm cluster. This further development can easily be observed in a purpose-built swarm-catching box, fitted on two opposite sides with closable ventilator grilles. When the two opposite sides are uncovered carefully a little at a time and some light is provided in the background, it is possible to see the silhouette of the mass of bees hanging like a bunch of grapes or in a heart shape.

3. THE SWARM

The swarm in the box is very quiet; only here and there can we hear a brief buzz or hum. If we knock gently on the box the bees answer with a brief, loud wave of hissing-humming that rises then falls in intensity: 'tzssschchschschchchsssss ....'

The beekeeper may leave this box undisturbed for up to three days. The stores that the bees have brought with them will only last that long. A swarm left alone undisturbed like this will lose about 100 to 200 g in weight in this period. Alternatively, a swarm may be hived the evening after it is taken.

Hiving the bees in a new home

In the meantime, the beekeeper prepares a new hive for the swarm. He cleans out the brood box and equips it with frames for supporting comb. These are made of strips of wood, usually referred to as 'bars,' forming a rectangular shape, longer in either the vertical or horizontal direction, depending on the type of hive that the beekeeper is using. They are hung vertically in the hive at equal distances from one another. The finished hive is closed up leaving the hive entrance open.

In the late afternoon of the hiving day the beekeeper collects his swarm from the cellar. He carefully opens the box and the swarm cluster is still to be seen hanging quietly. In front of the entrance of the new hive he has placed a horizontal board onto which he now

Swarm cluster in the box just before shaking it out [Photo: Tim Kraus, Kassel 1995]

Swarm being hived [Photo: Tim Kraus, 1995]

briskly shakes out the swarm. The mass of bees flows over one another like dough. A few bees immediately rise into the air again and fly round the area in lemniscatory curves; others begin fanning from their scent glands; but most of them creep in a uniform flow towards the hive entrance and steadily disappear through it into the hive.

Once again the queen is spotted as she is driven towards the hive entrance in the general flow. Other bees slowly take off, many start scenting in the way described and a subtle, difficult to describe, scent is noticeable in the immediate vicinity.

Meanwhile the beekeeper has carefully brushed the last few bees out of his swarm box. Three white structures are hanging parallel to

3. THE SWARM

each other the same distance apart on the upper wall of the box. They are heart-shaped round-oval and about two centimetres thick; the one in the middle is about twice as big as the ones at the sides. Closer examination reveals that the swarm has made three beautiful white combs with the characteristic hexagonal cells. Small white rods about two millimetres long can be made out on the bottoms of some of the cells. The beekeeper explains that these are eggs that have just been laid there by the queen.

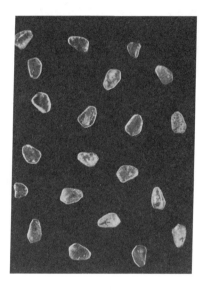

Wax scales, (enlarged 4×)
[From Zander and Weiss]

Then he points to the floorboard of the swarm box that he has put to one side. On it are a few dead bees, small twigs and a large number of rounded oval, white-opalescent platelets between one or two millimetres wide. The beekeeper says that these are scales of beeswax which the bees have 'sweated' out and which have fallen down during the construction of the comb.

He also points out that the spacing between the combs is 35 mm. He has hung his frames according to the same spacing so that the colony will build their comb in them. This allows him to move the combs individually later on when he is carrying out some manipulation in the care of his bees.

Comb with bee [Photo: Tim Kraus, Kassel 1996]

Meanwhile, after about fifteen minutes, most of the bees have disappeared into the hive. The beekeeper brushes the last few off the hiving board in front of the hive entrance and takes all his equipment away.

II. 'INSIDES'

4. Building Comb

The bee activities that will now be described take place inside. These processes can only be observed by the beekeeper if he opens the hive from time to time. The movability of the frames enables detailed observations to be made each time he does this. The accumulated scientific research in the specialist beekeeping literature enlarges the picture and thus gives a fair idea of the life of bees inside the hive.

However, the comb itself can only be seen if the bees have left it or been shaken or brushed off it. To study the hexagonal cell structure, a single comb is lifted out of the living whole.

How is comb created? Our first impressions are from the comb made when the swarm was in the swarm-catching box in the cellar, and from small round platelets that the beekeeper had described as wax particles 'sweated' by the bees. What is the explanation of this process?

On the underside of the worker bees' abdomen are glands arranged in four pairs from which the bees can secrete wax. For them to do this, certain conditions are necessary, as they cannot manage it alone.

Comb on a frame [Photo: Tim Kraus, Kassel 1996]

35

The conditions for wax production are:
— a sufficiently large colony of bees with a queen;
— an adequate food supply with carbohydrates (nectar, sugar) and protein (pollen);
— a time of year of increasing development (build up), or a swarm situation;
— enough bees with activated wax glands.

▲▼ 'Alfresco' [Photo: Klaus Bogen, Kassel 1994]

4. BUILDING COMB

Shaping space

The bees link together in a sheet by forming chains and hang from the lid or some other upper boundary (top bar of a comb frame etc.) in the cavity that they occupy. Increased heat production can be detected in the construction cluster (greater than 36°C); a number of bees are exuding wax scales. These are carried upwards with the legs and mandibles. The bees right at the top of the construction cluster knead the wax with their mandibles while at the same time probably mixing it with a secretion from one of their cephalic (head) glands (see diagram of glands on page 56). Then they attach it to the frame bar from below. The wax is first drawn into blobs. A little later they begin to shape it into a hexagonal matrix. By applying more wax to the edges of the developing comb its area is increased. As soon as this has reached a diameter such that, if rotated around its vertical axis, it would sweep out a sphere of about 5–7 cm, they start the construction of adjacent combs. This way the comb matrix grows more and more, eventually to hang in the space occupied by the swarm.

The development of the process can be experienced as a movement. The swarm cluster or the colony permeates a particular volume of the cavity in which they have made their home. Now the wax glands of a number of the bees are activated; wax is produced and exuded. Individual bees can manage this because they are part of a whole colony. The wax is taken up by the suspended cluster and moulded into its characteristic comb formation in an inner space.

Comb building stops when one of the above mentioned conditions are no longer fulfilled, when the comb as a whole fills the available space in the cavity, or when it has reached the size that, within the particular cavity, can be occupied and permeated by the formative capacities of the colony.

But what happens to a swarm that has found neither a suitable cavity nor has been 'taken' by a beekeeper? In such an emergency it builds its comb in the open on the branch of the tree on which it has settled.

Natural comb construction in a rear-access hive (a so called Baden three-storey hive) set up warm way by the swarm. [Photo: author]

Beehives

In what follows, in order to continue a more detailed study of the processes inside a bee colony, the box in which the beekeeper has hived the swarm has to be opened every day for a certain period. This requires that all the frames that already contain constructed comb be lifted out of the box. So we need first to insert a brief description of the various hive types.

There are basically two types of hive, those that the beekeeper can open from the top to lift out the frames, and those that are opened from the back, opposite the side with the entrance, thus allowing the frames to be taken out from the back (rear access hives).

Furthermore, the frames may be shorter than they are wide (broad format), in which case the bees build broad comb, or they may be taller than they are wide (tall format), in which case the

4. BUILDING COMB

bees build tall comb (rare in UK/USA). Frames are rarely, if ever, square.

The next distinction is over whether the frames are parallel to the front of the hive or at right angles to it. The first the beekeeper refers to as 'warm way' and the latter as 'cold way.' This terminology is connected with opinions on air flow in the beehive.

As regards the last mentioned distinction, an observer would notice clear differences in the construction of the comb by the swarm. With 'warm way' the swarm would build in the front frames hanging down parallel to the front wall of the hive. In doing so, the second and third combs would soon become the biggest and the comb at the front would at first remain smaller. The frames further back will gradually be built in too, those further from the front wall being started later. In rear access hives, the rear closure is usually a window which, with 'warm way' hives, allows these developments to be observed without opening the hive completely. With 'cold way' the swarm usually occupies the frames in the middle and from there builds the layers of comb to each side frame by frame. In this case the central comb is at first the biggest and those at the sides start off smaller. Only when the colony has built up sufficient strength to permeate the entire space do all frames gradually get filled with comb. Depending on the size of the swarm, this can take several weeks, or if it is being started at the beginning of summer it might not be completed at all that season.

This description may give the impression that observing a particular aspect (here the construction of the layers of comb) may actually only be grasped and presented selectively — that is, separated from other processes. The construction of the comb lasts for several weeks after hiving the swarm. Most of the job is completed in the first three weeks. But this process is inseparably connected with a host of other processes in the colony which, however, can also only be observed and described separately, so that a certain degree of comprehension is created from the outset.

5. Bee Development

The following description is based on inspecting a swarm in a top-opening hive. The frames are hanging 'cold way' and are tall format.

After the beekeeper has removed the hive roof, at least one more cover comes into view. It may be a wooden 'crown board' or a waxed cloth (the 'quilt') or — nowadays popular in Germany — a transparent sheet. Under this a gentle activity is detectable, possibly also visible. If the observer puts his hand on this surface he will be able to feel that it is warm in varying degrees in different places. Where most of the bees appear to be it is warmer than at the edges. The beekeeper slowly lifts up the sheet. A few bees come up on to the top bars of the frames and may fly away. A soft roaring increases from amid the general humming. The beekeeper gives a few puffs of smoke over the frames with his smoker, whereupon some of the bees go back down amongst the frames. 'Where there's smoke there's fire,' and animals generally flee from fire. Honey gatherers have been putting it to use since ancient times.

It is now the third day since the swarm has been hived and it is clear that the frames to the right and left are still free of comb. The swarm has suspended itself in the middle of about seven or eight frames. The beekeeper carefully removes the free frames and puts them to one side. Looking diagonally into the box from above, he can see the dark mass of bees that seem to be suspended in the same way as they were in the swarm-catching box. If he removes the next frame and the one after it, the bee cluster is split apart. He feels a certain resistance as he does this and points out how tightly the bees are holding on to each other and to the wood of the frame. The beginnings of a comb is visible in the frame next to the central one. In the middle frame, there is a comb about the size of a hand, and on the frame after it, there is also the beginnings of another.

By lifting the frames carefully, lots of bees can be seen hanging in chains on the lower margin of the comb. There are several vertical chains as well as some going diagonally or even almost horizontally to the side bars of the frame. Each bee clings to the next with its forelegs and hind legs like links in a chain. If we try to break such a chain with a stick or a finger there is a significant resistance to be overcome. However, almost all are broken when a frame is lifted out and most of the bees fall into the box.

5. BEE DEVELOPMENT

Egg laying

In some of the cells at the upper edge of the comb is a thick, gleaming, transparent yellowish fluid. The beekeeper tells us that this is either the honey reserves that the bees brought with them or freshly gathered nectar.

Furthermore, on closer examination, on both sides near the middle of the comb are quite a lot of adjacent cells containing eggs. These patches, about four centimetres wide, each comprise about fifty cells at the bottom of which are fixed little white rods. The beekeeper replaces the frame in the hive taking care to keep it in the same orientation and position as before.

At the next inspection two days later the observer notices that the area of comb has grown. The fourth and fifth combs have been started and the middle one is now seventeen centimetres tall and heart-shaped, somewhat longer than it is broad. The area of cells in which the queen has laid eggs is now at least as large as the palm of a hand and there are patches of eggs on both adjacent combs. A rough calculation shows that between 300 and 350 eggs have been laid on either side of the central comb and on the adjacent combs there are about 150 — that is, the queen has laid about 1000 eggs since she started.

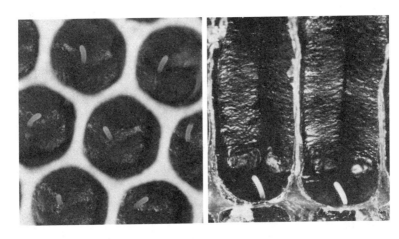

Eggs in cells
left: viewed from the entrance to the cell
right: cross-section view from the side   [From Zander and Weiss]

Laying queen surrounded by ring of nurse bees ('court')
[From Zander and Weiss]

During this inspection we notice the queen on a comb surrounded by a star-shaped ring of bees. Some are stroking her with their antennae. Others are offering their mouth parts and proboscises to her and she is taking something from them with her mouth.

The beekeeper says that the bees are constantly feeding their egg-laying 'mother' with a highly nutritious secretion, the so-called 'royal jelly,' from the hypopharyngeal glands in their head. The nutritional physiological analysis of this liquid shows that it is a mixture of easily digested carbohydrates and balanced amino acid compounds, or first-class proteins.

In the upper part of the comb there are significantly more stores of honey and nectar that have been deposited there in the previous two days.

Larvae

More discoveries are made at the next inspection a few days later. The middle comb has in the meantime grown so large that it has become attached to the side bars of the frame. In the cells in which

5. BEE DEVELOPMENT

the first eggs were laid are now small white curved grubs, or larvae, lying in a clear to whitish translucent liquid on the bottom of the cell. A liquid gleams here and there in the surrounding cells too, and on close inspection it is possible to see that here also the eggs have become tiny larvae.

As a rule, about 72 hours after an egg is laid by the queen a larva hatches out of it.

The liquid in which the young larvae float is almost identical to royal jelly.

The area of comb laid by the queen has grown and now covers parts of the fourth and fifth layers of comb. In the past two days she has laid about a further thousand eggs. Chains of comb-building bees have formed on the sixth and seventh frames.

On the fourth and fifth frames, which at this time form the outer combs of the developed nest, some cells have been filled with a dark substance. Many of them are deep yellow, others violet or black. In one of the cells there are two small orange clumps. The beekeeper explains that this is where the pollen baskets are being carried to. The flower pollen provides the protein component of the diet of the bees and their brood.

Three days later, the beekeeper has another look at what the swarm has developed so far in the hive. It has now been there for ten days. Comb has now appeared on the sixth and seventh frames and is already the size of a hand. Honey and pollen stores have already been deposited in it. On the fourth and fifth combs, there are already large patches of eggs, and the eggs that were visible three days ago have become small larvae. On the second and third combs the patches of brood have increased in size. The most recent and young larvae are surrounded on each side, and below, by a large area with eggs. In the meantime, these combs have also been fixed to the side bars of their frames. The middle comb has grown longer at the bottom and here too a broad ring of eggs surrounds an area of comb with larvae whose age increases towards the middle and the top.

Where eight days previously the first eggs had been laid there are now fat round grubs filling the diameter of their cells. The beekeeper calls these 'round larvae' and it looks as though they have hardly any more room to continue growing. The larvae in their cells do not all face the same way and the beekeeper says that they swim around in their food, always rotating round their tails in the direction they are facing. He adds that the older larvae are no longer fed with royal jelly from the

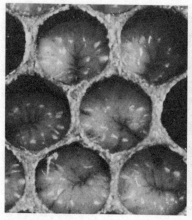 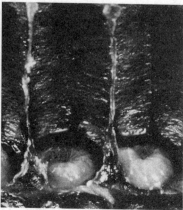

Round larvae [From Zander and Weiss]

cephalic glands of the 'nurse' bees — as those that feed and care for the brood are called — but with a mixture of pollen, honey and royal jelly.

As a rule, after three days as an egg, the young larva is fed for three days with royal jelly from the cephalic glands, and thereafter with pollen-honey mixture.

The beekeeper estimates that in the previous three days the queen has laid over two thousand eggs in new cells.

Pupation

The next inspection happens three days later. Vigorous foraging flight is under way at the hive entrance. The many pollen gatherers are conspicuous.

Inside the hive, the bees have not started to build any new combs. Where still possible, the existing seven combs have been extended to the sides of the frames and in all cases to the bottom. The stores of nectar and pollen have only grown a little. Obviously much of the food is now being used for feeding the 'open' brood.

In the combs at the sides, large patches of cells with eggs have appeared and the eggs that were visible three days previously have now become the youngest larvae.

The picture on the middle comb has changed a lot. The cells that eggs were seen in at the first inspection ten days previously are no longer open. Also the cells around them have acquired waxen,

5. BEE DEVELOPMENT

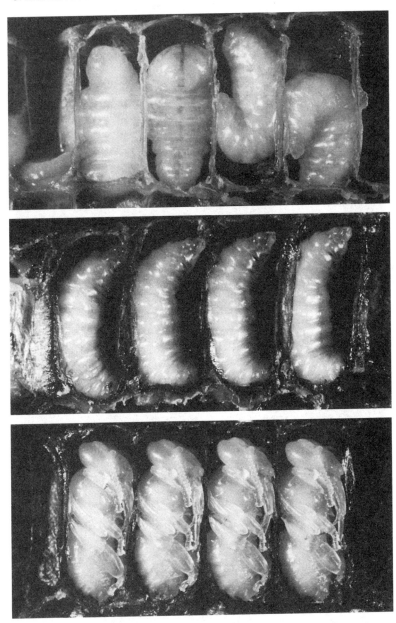

Brood development in a capped cell ('closed for redevelopment'):
Stretched larvae spinning cocoons (top)
Before pupation (middle)
Pupae (bottom) [From Zander and Weiss]

New comb with capped brood [Photo: Tim Kraus, Kassel 1996]

Wild comb with worker and drone cells [From Casaulta, Krieg and Spiess]

5. BEE DEVELOPMENT

slightly domed caps. Where that has not happened larvae are still visible, though they are no longer round and fat and lying on the bottom of the cell, but have straightened out towards the entrance to the cell and are almost looking out of it. The beekeeper says these more mature grubs are 'stretched larvae' and that the cells are capped by the bees nine days after egg-laying. During the next three days the larva spins a cocoon around itself and becomes a pupa that undergoes transformation into a bee. At the end of feeding, a stretched larva weighs about 0.12 g, which is at least five hundred times heavier than the egg that the queen laid nine days previously.

Drone brood

The next inspection occurs seven days later, on the twentieth day after hiving the swarm and the seventeenth day after the first inspection. The layers of comb have again grown bigger. Areas as big as a hand have now appeared in the eighth and ninth frames and the bees would probably have started building in the tenth and eleventh frames, but no chains of wax-making bees are to be seen there. The beekeeper points to this and says that a swarm builds vigorously for three weeks and then its construction productivity declines. This is dependent on the strength of the swarm and on the forage and weather conditions. Pollen has been deposited in the outer combs together with some honey. The middle three combs have now reached the bottom bars of their frames.

The picture has changed on the fourth comb. There are a few empty centimetres between it and the bottom bar but in the lower region of the comb the cells look different. The cells that have been moulded there have openings that are at least a third bigger in diameter than those we have seen hitherto. There are eggs at the bottom of these cells too.

The beekeeper points to a drone that is just crawling over these cells and tells us that in these larger cells the drones of the bee colony are being raised. They are bigger than the worker bees.

The brood nest has now grown bigger overall and in it almost all the brood that was present at the last inspection is now capped. Around it, up to the penultimate combs to the sides, are areas of brood with all the familiar developmental stages.

Bees hatching

We take another look at the colony four days later. There has not been any significant additional comb construction. The beekeeper carefully and smoothly moves the outer combs to one side in order to remove only the middle one. The picture has changed again. The area of cells in the middle of the upper part of the comb, with eggs laid by the queen twenty-two days before, were capped at the last inspection. These cells are now open again. Compared with the shining white comb at the beginning, they now appear tinged a brownish colour. The beekeeper points to this, explaining that this results from the cocoons that each hatching bee has left behind in the cell as a residue of its process of metamorphosis. These cocoons comprise the pupal silk, larval faeces and residues of the five moults (ecdyses) that the larva has undergone.

The beekeeper now points to the surrounding area of cells. A few cells are empty here too. Some bees are crawling around on them looking very different from the other bees. They look as if they are surrounded by a fine grey fur and their wings are somewhat creased and not yet unfolded. They extend their proboscises to other bees and are fed.

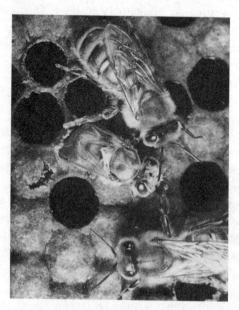

Hatching bees [From Zander and Weiss]

5. BEE DEVELOPMENT

The beekeeper points to a cell that is still capped. On the edge of the capping a small hole appears in which something is moving. Closer observation shows that it is half a mandible pincer sticking out and apparently gnawing the capping against the other mandible that is hidden from view. The hole gets bigger, the pincer is withdrawn and an antenna appears briefly, feeling about on the cell capping. The gnawing is resumed. The removal of the capping has proceeded further on another cell and the face of a bee is looking out of it. The antennae are feeling about and the head is extended outwards. Leg segments come into view and the round thorax is slowly squeezed out, apparently with a lot of effort. Once it is out, the abdomen quickly follows. The small, crumpled bee looks somewhat lost amongst the teeming activity on the comb. Shortly afterwards, it is offered food from the proboscis of another bee and takes it immediately.

The beekeeper tells us that a hatching bee is only fed if it has left the cell through its own efforts. They are not helped during hatching. If they cannot do it with their own strength they are killed and the corpse is removed from the hive.

The developmental stages of the worker brood are as follows:
— egg, 3 days;
— young larva, 3 days, fed with royal jelly;
— older larva, 3 days, fed with honey-pollen mixture;
— cell capped, 9 days;
— the round larva straightens into a stretched larva and spins a cocoon round itself up to the 12th day of development;
— after 12 days in the capped cell the young bee hatches on the 21st day after egg-laying by the queen.

Therefore the brood period lasts $3 + 6 + 12 = 21$ days.

The development of the queen and drones have different timings, which will be presented in the next section.

The timings for brood development are relatively inflexible. There is a constant temperature in the brood nest of the hive of about 35°C and an almost constant humidity of about 60%. Only when more extreme circumstances outside the hive disturb the internal conditions, might the development be increased by one or two days, or if the brood is damaged or dies.

A further inspection a few days later reveals that the area of hatched cells has increased. If the colony has enough forage, a portion of the vacated cells are filled with stores. The queen lays eggs again in the rest.

In trying to take in all these life processes in a mobile overview we might have the impression that the queen's laying activity sends an impulse of rhythmic regularity throughout the colony. We shall consider later whether there is any evidence in this impulse of movement in any direction.

6. Bee Life

The queen lays eggs in a spherical space of increasing size within the layers of comb that the swarm has produced in the first three weeks since it was hived. After a few days, and for a period of time, she lays over one thousand eggs every twenty-four hours in the newly created cells. The first young bees of the new colony emerge on the twenty-third day, after twenty-one days development. After that, more hatch each day. This is absolutely essential for the further development of the swarm. Lots of bees are lost daily during foraging activity. They fail to return to the hive from its surroundings because of age, exhaustion or other influences. As a result, the number of bees in the swarm decreases daily from the beginning.

But the hive must have sufficient bees to secure the essential life processes within it.

House bees — flying bees

In order to get a more accurate picture of these processes, they need to be split up for the purposes of observation. In their descriptions, people use concepts that are familiar to them from their own circumstances in life. Thus, in the colony there is an assortment of 'activities' of individual bees that can be observed and differentiated. These take place over a specific period, sometimes alone or in collaboration with other bees. Examples include guarding the hive entrance, building comb, exuding wax, feeding the brood, warming the brood nest, cleaning and 'clearing out.' There are many more of these activities.

If the observer were to watch individual bees of a colony over a period of many days, he would probably at first notice that the bees can be divided into two kinds: flying (or foraging) bees and 'house bees.' The latter are responsible for the maintenance of the life processes inside the hive, the former for the nectar and pollen gathering activities we have already discussed. It is possible to make this distinction so clearly because it is obvious from observing the bees that the flying (or field) bees do not take part in the work of the hive, but simply unload what they have brought in (nectar, pollen, water) in order soon to fly out again. As the number

Table: 1 Bee development. Course of brood events starting from a swarm situation

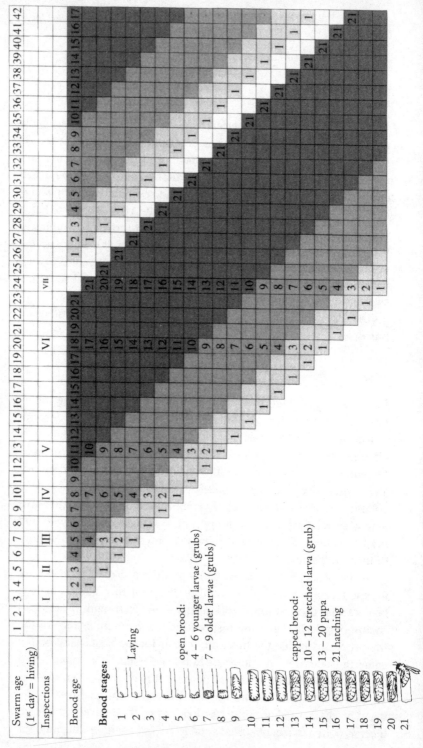

6. BEE LIFE

of flying bees needs to be kept relatively constant in relation to the strength of the colony, there needs to be a daily replenishment of those out hunting.

And in the hive there is a great variety of jobs that has to be constantly carried out. This raises the question of how the necessary division of labour arises.

The enquiring observer now conducts an experiment by carefully marking a single bee, or several of them, with a small coloured blob on the back of the thorax immediately after hatching. This enables him to trace where the bee goes subsequently.

Bee researchers make observation hives for this purpose. These contain only one large layer of comb covered on both sides by sheets of glass, which enables the activities of all the bees to be monitored. In Greece, 2,500 years ago, Aristotle observed the life of bees in this way and described what he saw. Instead of glass he used selenite (crystalline gypsum, calcium sulphate). Maurice Maeterlinck (1903) was also able to describe 'The life of bees' in this way.

Development from the inside ...

A newly hatched bee is first fed via its proboscis by other bees. Its movement over the comb continues unassisted. It almost staggers. Its hair appears grey, matted and ruffled. With hesitant movements at first the downy grey young bee cleans itself and smoothes out its still crumpled wings. It looks like a little, helpless child, but soon seems steadier and more resolved. After several feeds it begins to slip into recently vacated brood cells in its immediate vicinity and to busy itself in them. It gnaws the cell walls with its mandibles and licks them with its proboscis. In doing so it is apparently secreting something with which it polishes the cell wall until it looks smooth and shiny. This gets the cells ready again, either for the queen to lay in or to receive stores.

The young bees now feed from the cells of stores, especially on pollen (protein). Yet they remain close to where they hatched. By shivering movements of their wing musculature, which also has the effect of training it, they produce heat and thus help to maintain the brood nest temperature and maintain a constant 35°C.

Bees between two and three days old begin to feed the older larvae of the bee brood with a mixture of honey, pollen and royal jelly. Over and over again they stick their heads in the brood cells and do something there for a few seconds and then move on to the next. In between they regularly collect food from the store cells.

After a further two or three days the bees devote themselves to cells with younger and younger brood, busying themselves in caring for it, feeding it or inspecting it. At this age they also possibly join the bees that surround the laying queen for a time. The body of the mother of the hive is always being touched and licked. Bees are constantly offering her their proboscises so that she can take food from them.

In the period of life between the fifth, and ninth or tenth day of development it is also possible to see lots of bees taking on small reconstruction jobs on the cells or participating in capping them.

In the normal course of development, individual bees leave the hive for the first time around the ninth or tenth day. The young bees take their first flight to orientate themselves in the immediate surroundings of the hive. This is the 'orientation flight' that has already been described.

... to the outside

In the ensuing days we can observe the young bees taking the nectar from the incoming foragers and travelling to where the stores are being deposited. Hour after hour the fresh nectar, which at first is a watery liquid, is tended and thickened. To do this the bees take it up and expose it to the air of the hive. Over and over again it is sucked up and expelled. Each time a part of the water is evaporated in the warm, dry air of the hive. Then the honey is deposited in a storage cell to ripen further. At another time the bees 'stamp' with their heads into the cells in which the pollen baskets have been deposited. The pollen is compressed and stuck together with saliva and nectar. As a result it undergoes a lactic acid fermentation which breaks down its protein components and preserves it. Finally it is covered by a layer of honey.

Subsequently, the bees may join a comb-building cluster and there participate in exuding wax and moulding it into comb.

At about the eighteenth day of life, they may leave the building cluster and take part in cleaning and clearing out activities. They start 'guard duty' at the hive entrance, that is, they monitor incoming bees and try to drive away foreign insects. These can be strange bees from another colony, flies, moths or bumblebees. Frequently in late summer wasps try to get into the hive to steal the stores. It is often possible to

6. BEE LIFE

see two or three bees jumping on an approaching wasp, tangling with it and trying to sting it. Another job at the hive entrance is ventilation. The bees hold on tight with their feet, many standing in formation one behind the other, and move their wings powerfully thus blowing a steady flow of air out of the hive. This airflow is almost hot and humid, especially on mild summer evenings after a particularly successful day's foraging, and it smells sweetly aromatic.

The bees stay a further two or three days round the hive entrance until they are about 20 or 21 days old. Then they become foragers. When, at the height of summer, the development of the colony is still in progress they remain as flying bees for between 10 and 20 days more until they die somewhere in the surroundings.

Metamorphosis of the glands

Scientists have investigated bees using lots of different methods and discovered that the course of development of these externally observable activities is accompanied by changes in physiological processes within the bees. They have various glandular systems within them which gradually develop, reach maximum secretion, and then regress or change their function.

In the sequence of the processes described the following glands are active:
— thoracic salivary gland (cleaning cells);
— royal jelly glands in the head (feeding older brood, young brood and the queen);
— wax glands (comb building);
— scent glands (fanning, scenting);
— venom gland (filling the venom or poison sac).

During further development the royal jelly glands, together with the thoracic salivary glands, supply the enzymes with which the bees refine the gathered nectar into honey, and they remain active to a lesser extent in foragers.

It should be emphasized that our attempt here to describe the development of the bee from hatchling to flying bee is statistically ideal, at least as regards the temporal components. In contrast to the development times for the individual brood stages from egg to hatched bee, which are fixed to within a few hours, their further development is more subject to the inner requirements of the overall

Schematic representation of the glandular system of a honey bee:

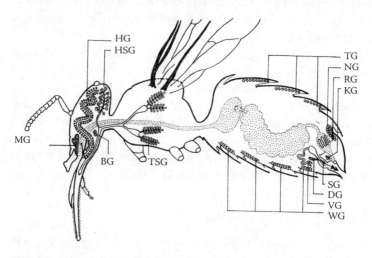

Cephalic or head glands:
HG = hypopharyngeal (royal jelly) gland
MG = mandibular gland
HSG = head salivary gland (head labial gland)
BG = buccal gland

Thoracic glands:
TSG = thoracic salivary gland (thoracic labial gland)

[From Casaulta, Krieg and Spiess]

Abdominal glands:
DG = alkaline or Dufour's gland
NG = scent or Nasonov gland
VG = large, acid or poison or venom gland
RG = rectal gland or pads (rectal papillae)
TG = tergite glands (Renner Baumann glands)
KG = sting chamber or Koschevnikov's plate gland
SG = sting sheath gland
WG = wax gland

colony organism, and as a result varies to a large extent. Certain developmental stages may be completely omitted if in the hive there is at the time no, or only limited, need for them. Thus, for example, the wax glands do not reach full productivity in all bees when the space available for comb-building is filled and only capping or remedial work needs to be done. Also, when vigorous foraging is in progress as a result of favourable conditions, the house bees, that have not gone through all the stages, become flying bees. Conversely, flying bees may, if necessary, reactivate their wax or food glands and resume the work of house bees to an extent that is sufficient for the development of the colony.

6. BEE LIFE

Observation for a longer period shows that this aspect of the life of honey bees is only partially subject to an inner principle, and involves an active and reactive relationship with outer conditions and requirements.

Interpenetration

In considering the overall development of single bees amongst all the bees of the same age, another movement can be discerned. At first the bees stay in the hive close to where they hatched. Then they tend to move further and further away from that place, gradually going through all the activities of looking after the brood. Then comes the orientation flight of the young bees after which they are more active in the outer region of the inside of the hive. For example, managing the stores and depositing honey and pollen stops at the margins of the brood nest, and building generally takes place on the margins of the existing comb. Cleaning, ventilation and guarding the hive entrance take the bees more and more towards the outside. Then, as flying bees, they repeatedly 'disappear' into the infinitude of the surroundings, and eventually even stay there for ever by physically passing away.

Thus, in the hive we can experience two meeting and interpenetrating movements. From one direction comes the flow of substances that are constantly being brought in and there internalized. A more selective passage through the hive is observable in the queen. She is constantly being tended and fed and constantly producing new eggs which she lays, on the one hand, point-shaped in each cell, and on the other, rhythmically and spherically distributed in space. The transformation of her activity into the enormous metabolic output of the feeding bees and the developing brood shows another quality.

The metamorphosis proceeds in the other direction in the growing — initially more inwardly orientated, later more outwardly developing — regions of activity of the worker bees in the hive, and ultimately in the physically almost dispersing gesture of their spreading out into the surroundings of the hive.

7. Summer Impressions

The air is filled with the sound of buzzing.

Listening with closed eyes gives the impression of standing under a great bell that is constantly being firmly rubbed. The whole surroundings of the apiary are humming. We remember the experience of the swarming bees. However, this time the bulk of the bees are not flying around looking for something, but going directly to a place that is out of sight. Bees hurry almost incessantly from every hive entrance; they take off without delay and fly away quickly. Meanwhile, others descend as points of light out of the blue sky, briefly hovering in front of the hives before disappearing through one of the entrances. There is something overwhelming about this humming sound. The activity of several thousands produces a feeling of elation on this hot, sunny day. The visitor feels something like an inner rejoicing.

'There is a flow on,' say the beekeepers of such a phenomenon, when apparently everything that can fly rushes into the surroundings and returns again fully laden with nectar. This is likely to happen when round midsummer the lime trees are in flower or when the trees are producing honeydew; that is, when the leaf aphids in oak, sycamore or lime, or when *Lachnidae* in pine or spruce, secrete a sweet clear liquid which drips on the leaves or needles and forms a shiny sticky film.

This is honeydew. Many insects, midges, hover-flies, wasps, ants and, of course, bees come to it to lick it or suck it up.

About seven weeks have passed since the swarm under observation was hived in its new box.

The queen has already twice laid up the comb that in the meantime has grown further. In doing so, on some days she has probably laid more than 1,500 eggs in newly constructed cells or in cells vacated by hatching bees. On such days she lays at least one egg a minute and the total laid in 24 hours together weighs more than the queen herself. In her large abdomen the ovaries comprise up to three hundred strands, the ovarioles, in which new eggs are continually being formed and matured one behind the other. They then slide along the oviduct into the vagina. The secretory duct of the spermatheca has its outlet there. In the

7. SUMMER IMPRESSIONS

spermatheca is the sperm store which is filled with between five and seven million spermatozoa after the mating with drones on the nuptial flight. This store usually suffices for the whole life of the queen, which can last for between three and five years. How the sperm stay alive and undiminished in their fertility for so long is a question that has so far not been answered by biologists researching the matter.

When an egg passes the outlet of the spermatheca, sperm are released, but only if it is being laid in a worker or queen cell. Drone cells are laid in without fertilization of the eggs. This means that drones develop from unfertilized eggs and are thus 'fatherless.' This process is called 'parthenogenesis.' It is relatively rare in the natural world and thus the production of drones amongst bees is a biological curiosity.

It is possible to make a number of new observations if the hive is inspected around this time. The first thing to notice is that there is now comb in all the frames. Although comb construction in the frames at the sides is not complete, honey is already being stored in the outer combs together with freshly brought in nectar, and on the adjacent combs there is already some brood. All brood stages are well represented. Above it fresh nectar and pollen is being stored and, above this, the comb has been drawn out wide and contains a ring of honey as long term storage.

Bee dances

The house bees are busy everywhere with the activities already described, but some of them are noticeable for their special behaviour. They move round the comb in a pattern that always returns to the same point. They waggle their abdomens rapidly from side to side as they move in a short straight line then return to the starting point in a semicircle only to repeat the 'wag-tail' dance along the straight line, this time completing the other half of the circle on the return, and so on. There are quite a lot of bees showing this behaviour, especially in the lower part of the comb. Many are carrying pollen baskets, others not. Some are walking vertically downwards on the comb, others upwards, many horizontally to the left or right and some diagonally upwards or downwards. But all of them keep to their own direction and exhibit this waggling movement of the abdomen as they walk along the straight line. Some trace out the pattern of movements more vigorously, quicker and more frequently; others more deliberately, slower and less frequently. The 'dancers' often repeat their figure of eight for several minutes without a break.

Quite a lot of bees on the comb are paying attention to them, touching them with their antennae, following them briefly and then going away again. This movement behaviour is reminiscent of that experienced on a hanging swarm until it eventually flies to a new site or a new hive. Bee scientists call this behaviour in the hive 'neural communication.'

Karl von Frisch (1895–1981) studied these phenomena in a long series of experiments over several decades. He discovered that bee dances are connected with a particular foraging site found by a bee on one of her flights, and he observed that after she has returned to the hive more bees from the hive soon seek out the same site.

Frisch did experiments with dishes of syrup placed at various distances, and in different directions, from his observation hives. He marked the returning bees with coloured identification dots so that he could observe their behaviour in detail on their return to the hive. Thus he was able to discover that the bees that had discovered a source of forage, on return to the hive stimulated other bees with the dance described to fly to the same source. And he succeeded in decoding the dance pattern.

There are two distinct components to the dance. One is the direction of the straight line on the comb. The other is the frequency and intensity with which this distance is traversed with the wag-tail dance.

Bees sense in flight the direction from which the light of the sun is coming. As bee eyes can detect polarized (orientated) light from the sun it is possible for them to find the way at times when they cannot actually see the sun itself and as long as there is sufficient daylight.

As they walk on the comb in the darkness of the hive, the bees sense the direction of gravity. If the dancer moves vertically downwards in the gravitational field, the dance stimulates the attendant bees to fly away from the sun when outside, into the darkness so to speak. If the dancer moves upwards against gravity, it stimulates the bees to fly towards the sun into the light. Upwards to the right and downwards to the left mean fly to the right in relation to the sun. In this way every angle of flying direction in relation to the light source can be danced on the comb.

7. SUMMER IMPRESSIONS

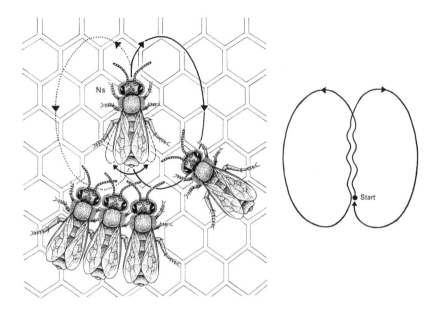

Pattern of the waggle dance on the comb [From Casaulta, Krieg and Spiess]

With dancers that sometimes dance for hours to stimulate bees to fly to a particular source of forage, the direction of the dance on the comb changes with the position of the sun in the sky, without the bees having to leave the hive. If a bee is dancing, for example, vertically upwards in the morning, which directs flight towards the sun, at noon it dances to the left on the comb — that is, lightness/lighter to the right, heaviness/darker to the left — and directs flight at right angles to the sun's position with the sun on the right. In the evening it would dance downwards in the direction of gravity or heaviness, and thus direct flight away from the sun.

In watching the second visible component of the waggle dance, we noticed that different bees cover the straight course through the dance pattern at different frequencies. And it can be observed that a bee, with its foraging site closer to the beehive, completes her passes through the dance more frequently than a bee whose forage site is further away. Thus, the attendant bees are stimulated not only to fly in a particular direction but also they receive the information as to how far they should fly in this direction.

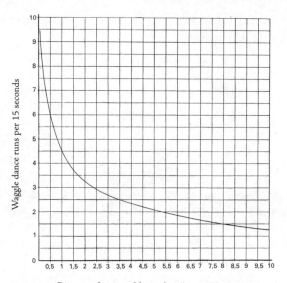

Frequency of waggle dance cycle versus distance to the source of forage
[From Casaulta, Krieg and Spiess]

Finally, the dancing bees also carry with them the scent and taste of the forage source, and even the sugar content of the nectar and the yield of the source is conveyed by the vigour of the waggle.

If the observer creates for himself a picture of the circumstances of the bee dance, he may recognize that with it bees can immediately convey to other bees in the hive their experience of discovering a source of forage, as well as where it is in the surroundings. The stimulus for the outbound flight is so exact that bees setting off for the first time do not have to look for the forage site. We might have the impression that they know where it is when they set off.

With respect to the swarm suspended out in the open it should be noted that in this case the dances described stimulate the bees to visit the cavities that have been found, in order to assess their suitability as new accommodation for the swarm. The swarm then flies away from its starting place when a large proportion of its bees have danced the dance for a new site and at the same time have got to know it.

7. SUMMER IMPRESSIONS

To summarize once more the visible elements of the waggle dance and what they mean:

Direction of movement on the comb	stimulates	Direction of flight
in the direction of gravity, heaviness	—	towards the darkness, away from light
in the direction of lightness, away from gravity	—	towards the light, away from darkness
Frequency of waggle dance cycle	conveys	Flight distance
higher frequency	—	closer to the hive
lower frequency	—	further into the surroundings

Here it may appear as if the terms 'closer' or 'further' are expressing something about the sensations of the bees in relation to the rest. Viewed from the centre of the hive, the immediate spheres may be sensed as 'closer together' and the distant spheres as 'further apart.' Is this expressed in the frequency of the waggle dance?

III. DEVELOPMENTS

Apart from the phenomenon of dancing bees, further inspection of the swarm around this time reveals several combs of the drone cells discussed previously. The drones have in the meantime hatched out of the cells that had just been laid at that time and these have recently been laid in again. Others are still capped. The capping curves upwards above each cell like a dome so that the drone comb area looks noticeably thicker than worker brood.

Here we need to mention the development cycle of drones in comparison to that of normal worker bees. The latter follow the pattern: egg, 3 days; young larva, 3 days; older larva, 3 days; cell capped at 9 days; 12 days of metamorphosis; hatching on the 21st day after egg-laying.

With drones the timing is as follows: egg, 3 days; larva, 6 to 6.5 days; cell capped on the 10th day; at least 14 days of metamorphosis. This means that drones hatch at least 24 days after egg-laying, although it could be 25 or 26 days. The entire sperm supply is fully developed by the round and stretched larva stages between the 8th and 14th days. Nevertheless, a drone is not mature enough to fertilize a virgin queen until the 10th to 14th day after hatching.

8. Queen Cells: Queen Development

The next interesting discovery at this inspection are two suspended, longish cone-like outgrowths at the edge of the lower third of a comb. They are not quite as thick as a little finger and are about 3 cm long. Their surfaces are covered with small depressions and look as if they have a *hammered* finish. In the bottom of one of them is a round opening into which a bee has just slipped almost completely, to the extent that only a part of her abdomen is still protruding. The other cone is capped with a small dome of wax. Now the opening of the first has been vacated by the bee and a light, gleaming white mass is visible in which a round larva is lying. It is comparable with a 4–5 day old worker larva.

The beekeeper explains that these are queen cells in which young queens are being reared. Clearly the colony is about to 'supersede' its queen. The present queen, who is now three years old, is failing. The colony senses this and sets about replacing the old queen.

A queen bee constantly secretes a pheromone — a hormone-like scented substance — through special glands that are situated predominantly in her head. This is constantly received by the attendant bees and distributed throughout the hive via the uninterrupted social

Queen cups (left) and a capped queen cell (right)
[Photo: Günther Friedmann, 1998]

8. QUEEN CELLS: QUEEN DEVELOPMENT

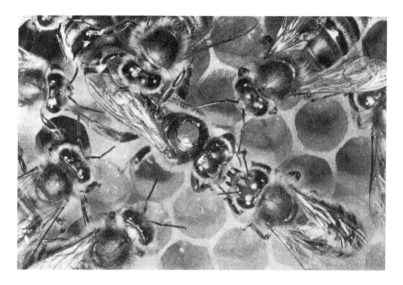

While she is being tended the queen is constantly being sensed by the workers [From Doering and Hornsmann]

exchange of food that takes place between all the bees. If this subtle flow of queen substance decreases in strength, it usually leads to profound changes in the further development of the colony, for example, to the aforementioned supersedure.

If the queen is lost for some reason, for instance, through a clumsy manoeuvre by the beekeeper, or if she is removed, the bees realize this within half an hour, as a result of the interruption of the pheromone flow. They sense that they are queenless. They run all over the place

Social feeding between two worker bees [From Zander and Weiss]

looking for her and soon begin to roar loudly. Beekeepers refer to this as 'moaning.' It also happens in a queenless colony for a short time after someone has knocked on the outside of the hive.

If such a colony has access to open worker brood, some of the cells are drawn out into 'emergency queen cells' which are similar to normal queen cells. The younger the larvae in these cells the better the emergency queen that is produced from them by the bees. The older the larvae — that is, beyond three days — the smaller the number of ovarioles which will reach full development and, therefore, the lower the laying capacity of the queen that results. This is connected to nutrition. The older worker bees are, of course, fed with a mixture of honey, pollen and a small amount of royal jelly. This ultimately turns (moulds) them into worker bees through the development of the appropriate glands and gathering organs, and through suppression of development of the specific female reproductive organs.

By contrast, a larva developing in a normal queen cell is supplied with large amounts of royal jelly, the white substance seen covering the whole cell 'floor' (ceiling in the hanging cell) and in which the larva lies (hangs).

Besides the feeding there is another factor to take into consideration, which distinguishes the development of the worker from that

Emergency queen cells are drawn over cells with open worker brood.
(From Zander and Weiss)

8. QUEEN CELLS: QUEEN DEVELOPMENT

of the queen. Worker bees grow and develop over twenty-one days in hexagonal cells in the body of the comb. Queen cells are rounded from the outset. A queen cup, usually situated on the edge of the comb, is at first almost spherical with an opening at the bottom. Then it is 'expanded.' If the queen lays in it and the lava develops then it is gradually drawn out into a cylinder and ultimately capped with a hemispherical dome at the bottom. With all these differences it is hardly surprising that the period of development of a young queen has its own rhythm.

After three days of the egg stage and five days of the larval stage the cell is usually capped and after eight days of metamorphosis the queen emerges; that is, sixteen days after the egg was laid.

Age in days	Queen Beekeeping term	Moults (ecdyses)		Age in days	Worker Beekeeping term	Moults (ecdyses)	Age in days	Drone Beekeeping term	Moults (ecdyses)	
1				1			1			
2	Egg			2	Egg		2	Egg		
3		hatching		3		hatching	3		hatching	
4		1st moult	open queen cell	4		1st moult	4		1st moult	open brood
5	Round larva (or grub)	2nd moult		5	Round larva (or grub)	2nd moult	5		2nd moult	
6		3rd moult		6		3rd moult	6	Round larva (or grub)	3rd moult	
7	Stretched larva	4th moult (capping)		7		4th moult (capping)	7		4th moult	
8				8	Stretched larva		8			
9				9			9		(capping)	
10	Pre-pupa			10			10	Stretched larva		
11		5th moult		11			11			
12	Pupa (nymph)		capped queen cell	12	Pre-pupa		12			
13				13		5th moult	13	Pre-pupa		
14				14			14		5th moult	
15		6th moult (hatching)		15			15			
16	Queen			16	Pupa (nymph)		16			capped brood
				17			17	Pupa (nymph)		
				18			18			
				19			19			
				20		6th moult (hatching)	20			
				21	Worker		21			
							22		6th moult (hatching)	
							23			
							24	Drone		

Table 2: Overview of the brood developmental stages of queen, worker and drone

As a rule, most of the queen cells are drawn in a colony that is in swarming mood. Frequently, between one and two dozen larvae are developed into young queens. From a biological point of view, in strong, healthy and vigorous colonies, swarming is the method by which the colony reproduces, rejuvenates and spread itself further. A 'prime swarm' with a queen that has been active in the colony until that point will normally emerge on the day that the first 'swarm cell' (queen cell produced with the bees in swarming mood) is capped, unless it is delayed or prevented by unfavourable outer or inner conditions such as bad weather, forage limitation or a queen that cannot fly. In other words it emerges on the ninth day after the queen has laid an egg in the first queen cell. A week later, with the subsequent hatching of the virgin queens, several secondary or after-swarms can occur, often referred to as 'casts.' These may continue until the inner conditions of the colony bring it to an end through the killing of the unhatched young queens. This may be carried out by the last queen to hatch, who bites into the remaining cells, or by worker bees. The residue of the colony has at its disposal all the comb with the stores it contains. The number of bees in it is replenished by the hatching brood in the ensuing days. About five days after hatching the young queen flies once, or several times, to the drone congregation area to be mated. But she only starts to lay when the last brood from the old queen has hatched.

9. Maturity

The next inspection of our swarm, that has now become a strong colony, takes place about two and a half weeks later. Midsummer has passed. The lime flowers are fading and the woodland forage is drying up. Despite the good weather, the foraging traffic is rather indifferent. 'Beards' of bees are hanging in front of the entrances or from the alighting boards of some hives. These are bees that have nothing to do in the hive and who would produce too much heat in there, and yet cannot find any forage worth bringing in. Some bees are flying noticeably restlessly, as if looking for something around the hive, and hover up and down by the bees in front of the hive entrance. Sometimes they try to get into a hive but are immediately driven away by the bees at the entrance. And wasps are seen behaving in the same way.

The bees doing this are called 'robbers' by beekeepers. They have become redundant in their own colonies because of the shortage of forage. Now they are trying to obtain from other colonies what they cannot find in the environs. Unfortunately, if a colony is not strong enough to defend itself it will be robbed clean within a few days and collapse. In order not to provoke robbing, inspections at such times last only a very short time. Anything needed for knowing what measures are still to be taken for late summer management must be completed in two or three minutes.

The outer frames are now fully built out with comb and these are filled from top to bottom with honey. The cells are already largely capped with wax. This tells the beekeeper that the honey is ripe and can thus be stored for a long time. The two full outer combs on the right and left hold about 2.5 kg honey. The beekeeper could now extract this but he prefers not to interrupt colony development. Also, by leaving it to the bees he is spared having to supplement their winter stores by feeding them sugar, or at least it reduces how much they have to be fed.

The comb with the queen cells is carefully removed. It is immediately clear that one of the two queens has probably hatched as the bottom opening of the cell is round and open and the cell is otherwise intact. The other is still closed at the bottom but it has a big hole in the side through which it can be seen that the cell is empty. It has probably been chewed out.

A queen is walking about on the adjacent comb. The beekeeper thinks that she is probably the young queen, as he had marked the other with a coloured tag on the back of her thorax. The new queen looks very large by comparison with the workers round her and is obviously already laying. She has just stuck her head in a cell and now is turning round and introducing her abdomen into it. She dips her body backwards into the cell up to the middle segment, steadying herself sideways with her reddish-brown glistening legs. A moment later she rises and wanders off in search of another cell to inspect. As she does this she is constantly surrounded by a circle of bees who look after her and feed her. A longish, white egg can be seen at the bottom of the cell she has just left. The keeper carefully replaces the comb in the hive and inspects another closer to the outside. He is trying to get an overview of the colony's stores.

This comb has a broad ring of honey round the brood nest. Somewhat surprised, the beekeeper points out a bee with a coloured tag on its back. He says that this is the old queen and he gently places the comb back into the box. He has seen enough and closes up the hive carefully. It had already become somewhat noisy round the hive as a few robbers were trying to gain entry from above.

The beekeeper explains that normally there can be only one queen in a colony. A colony would usually be able to support only one queen or one of the queens would go on trying to kill the other. Only in colonies which supersede the queen might it happen that the old queen carries on her activities for a while beside the young one. But it is rare that beekeepers have the opportunity to see two queens in a colony. In this case he hopes that he has not interfered with its development by carrying out this inspection.

Reduction — departure of the drones

Visitors to an apiary after midsummer may observe something else. It is early morning. The dew is on the grass and the summer sun is climbing over the horizon above the mist. The birds are still singing but are somewhat more muted than a couple of weeks before. Small puddles of condensation lie on the alighting boards of the hives as evidence of the processing of the nectar during the preceding night and the vapour from the air being blown out of the hive.

A bee drags a longish-round white object out of the hive and lets it drop from the alighting board. A lot of these white bodies, each

9. MATURITY

Drone being dismembered by a wasp [From Zander and Weiss]

about as thick as a pencil and at least a centimetre long, can be seen on the ground in front of the hives. Closer inspection shows that they have parts which are vaguely suggestive of bees: a head with eyes, antennae attached, mouth parts, bodies with folded-in leg segments, abdomen with light rings ending bluntly. But they are all colourless, somewhat translucently white, occasionally light-brownish. Some of them have eyes tinged with violet. The beekeeper recognizes this as drone brood that has been cleared out of the combs by the bees. A few round wax cappings scattered around suggest that the bees have uncapped the brood, pulled out the motionless pupae and carried them out of the hive.

Around noon on this day the drone traffic is noticeably less than it was a fortnight earlier despite the sunshine and hot weather. Some of the colonies have no more drones flying whereas others have only one now and again. They are driven away from the hive entrance by the workers if they try to get in. The guard bees send them away, push them with their bodies, and, with their mandibles, tear the wings and legs of these ponderously crawling, big-eyed droners. Now, hardly a single drone manages to gain entry into any hive whose bees are showing this behaviour. Many fly off to another hive and try to slip in there. Others remain sitting below the alighting board where in the cool of the ensuing night they become moribund and drop to the ground. Tits, sparrows and other birds help themselves to the pupae and fly off with them. Wasps fly round the hives and dismember the defenceless drones with their mandibles. The head, legs, wings and abdomen are separated from the thorax which is then carried away.

After a few days there are no more drones flying from this apiary into its surroundings. The beekeeper takes this as an indication that the colonies there are 'queen right.' In other words, at this time of year they can sense that they are inherently harmonious, each with a vigorous queen at the centre. If a colony still has some drones in late summer or is still tending drone brood, the beekeeper can assume that it intends to supersede because, for whatever reason, it is not satisfied with its queen. A colony in Central Europe starts raising drones at about the beginning of spring. Thus, the first drones hatch round Easter and are able to fly ready to mate at the end of April or beginning of May. In early localities, hives may throw prime swarms at this time. By the end of July or beginning of August, most of the drones have disappeared again and in Central Europe swarms are not normally expected after that time.

Wintering

A few weeks later, the heat and drought of summer is well past its peak; the grain-fields have been harvested and are being prepared for the next crop; the clusters of berries on the rowans are bright red; the first apples are already rotting under the trees and the wasps are flying close to the ground looking for food. They will find the smallest holes to get into the hives. The beekeeper has reduced the size of some of the hive entrances to make it easier for the bees to defend themselves against any wasps that seem intent on robbing. Yet for a while in the often slightly misty, humid coolness of daybreak they can penetrate the hive unchallenged. However, as they usually quickly come out again and fly onwards to hunt elsewhere, and do not hurry directly back to their nest, their success is probably somewhat limited.

Nevertheless, any dead bees and other debris thrown out of the hive is torn apart and carried away by ants and wasps.

Even on hot days the bee traffic in the apiary is rather moderate and subdued. The bee beards, often seen at hive entrances at the end of July, have now disappeared. A look in the swarm colony that we have been following shows that the number of combs with brood has decreased. It is still to be found on about 5–6 combs. The brood area has decreased too and at least three quarters of it is capped. From this it can be concluded that in recent days the queen has been laying fewer and fewer eggs in the cells and consequently fewer replacement bees than before will develop from these.

9. MATURITY

The outer combs are now only visited by a few bees and are largely filled with capped honey. Even over the brood areas there are wide rings of honey, mostly capped. The brood is surrounded with a ring of cells filled with pollen and some of these are covered with glistening honey.

The beekeeper makes another survey of the colony's stores and finds them adequate for over-wintering. This colony does not need supplementary feeding with syrup. This is not always to be taken for granted. Without the support of artificial feeding many swarms do not manage with their own resources, either to build sufficient comb or to produce and care for enough brood to create a colony strong enough to over-winter, or even to bring in sufficient stores to continue. The colony has to succeed in bringing about a balanced relationship of comb, brood and stores. For one thing, this is dependent on outer conditions of the apiary site such as microclimate and forage development, as well as on weather outcome. For another, it is influenced by a number of inherited traits as well as the individual strength that it can summon up out of the act of swarming.

In order to follow the further development of the swarmed colony, the next inspection may take place round Michaelmas. By then the leaves are beginning to change colour and the October sun creates a

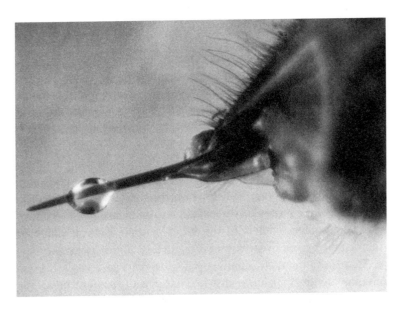

Bee sting with droplet of venom [From Doering and Hornsmann]

Comb with winter bees below capped honey stores [From Zander and Weiss]

multitude of warm, earthy tones in the surroundings. Many days still seem quite mild, but the nights are noticeably cold. The dew stays on the plants late into the morning. Spider webs shimmer in the morning sunshine which slowly disperses the light ground mist lying in the hollows. Even in the warm hours around noon, there is only weak foraging traffic. Pollen gatherers are in evidence on Michaelmas daisies and goldenrod. Otherwise there is merely a brief orientation flight of young bees.

In the colony, the outer store combs are almost abandoned. A thick coat of bees is now gathered on, and under, the combs where before we saw the brood nest. The bee cluster is suspended between the seams of comb and hangs down almost to the floor of the hive. A carefully removed comb reveals closely packed bees almost in two layers interleaved like roof tiles with their abdomens outwards, pointing downwards and their heads and thoraxes pointing inwards and upwards. This coat of bees is only moving very slightly and the colony flares up a bit on being disturbed. A few bees point their abdomens in the air and the outstretched stings can be clearly seen with tiny droplets as clear as water on the ends of them. This is a small amount of bee venom.

There is no more brood, or at most small patches of capped brood cells, on one or two combs at the centre of the cluster of bees. The remaining young bees will soon hatch from these. This shows that the queen laid the last cells around the equinox.

9. MATURITY

A colony's winter cluster forms over the vacated area of cells of the previous brood by hanging in the gaps between the combs close to the hive entrance. The uppermost bees of the cluster occupy the lower margin of the ring of honey of the comb they are on. A colony in this condition is ready for winter.

Unless there are special circumstances, the beekeeper does not need to open the hives again before the following Easter. The entrances are restricted or fitted with protective grilles (mouse guards) to stop mice getting in. The keeper carefully slides a sampling board into the floor of the hive under the bee cluster. By examining what falls from the over-wintering cluster onto this board he can draw conclusions as to the status and condition of the colony without having to open the hive.

10. Winter Inspections

The mood at the apiary even in the daytime is a damp, silent gloom. The leafless branches of the trees and shrubs look wet and slippery. When it is not raining, landscape and sky are frequently obscured by fog. The sun rarely shines directly on to the ground. Birds only occasionally fly past. At most we notice a brief flitting amongst the branches and a soft piping or short twitter. Occasionally, a tit comes to the apiary and quickly picks up a bee that is lying torpid on the alighting board or on the ground.

It has to be very mild with the air temperature over 12°C to tempt at least a few bees out of the hive entrances. These fly round the hive as the young bees do in their orientation flight. Some of them visibly void their recta as a spray, or deposit a brown blob on the front of the hive. In a healthy colony, bees never defecate inside the hive. Before the anus is a greatly expandable rectum in which indigestible residues of food and excretion products of metabolism gather until a bee can make a voiding, or cleansing, flight to deposit the faeces in the surroundings. The volume of the rectum is normally sufficient even for the long period of winter when the bees cannot fly out because of low temperatures, which can often last for three months or more.

When a beekeeper visits his apiary in November, he generally only checks externally whether everything seems in order. The hives are not opened and, if there is nothing untoward or noticeable, the visit is rather short. Perhaps he withdraws the sampling board that was inserted in the floor at the last visit. Usually a quick inspection of it suffices and it is shaken or scraped clean of deposits and reinserted. The sampling board comprises a water-resistant sheet inserted below a fine mesh (open mesh floor system), which is also sometimes designed for removal without opening the hive. There are a few bees lying on the mesh, mostly hunched up with their legs drawn in. These are the bees left from the summer that have reached the end of their life-span during the preceding days. The beekeeper lifts out the mesh and shakes the bees onto the grass.

10. WINTER INSPECTIONS

Reading the debris

There are some strips on the sampling board, the outer ones thinner and shorter, the inner ones longer and broader. They appear to form a circle. Close inspection shows that these strips are made up mostly of fine brown crumbs. Rubbed between the fingers they feel soft. They are crumbs of wax. The beekeeper explains that the bees form a broad sphere between the combs and this is called the winter cluster.

The bees in the upper half of the cluster gnaw away the cappings of the honey cells to get at their stores. The crumbs of wax fall down the gaps between the combs. This gives rise to the strips already described. The size of a cluster between the combs is indicated by the number of strips on the sampling board.

The beekeeper calls what has fallen down 'debris.' There are also other bits in this debris and the experienced beekeeper can draw several conclusions from them regarding the condition of the colony without having to open the hive.

Dry cappings without significant amounts of other components is a good sign. Black crumbs and cocoon indicate presence of wax moth larvae living in the comb. Chewed bees indicate that a shrew has visited the hive. The musculature of a bee thorax is certainly a tasty meal for it. Mouse faeces and nest material (dried grass, leaves) indicate that a field mouse has nested in the hive and has helped itself to bees and honey. Generally, moist debris indicates that the colony is still raising brood. This increases metabolism and thus creates more water vapour. In that case, there are often pieces of dead bee brood in the debris that the bees have pulled out of the cells.

In Central European latitudes, beekeepers like to have brood-free colonies by November. Brood needs a constant temperature of 35°C, which has to be maintained by the bees against very low external temperatures. The brood also has to be fed. As a result, the bees do not cluster so tightly and motionlessly. This increases the consumption of stores and, furthermore, has an unfavourable effect on the climate inside the hive because of the associated increase in humidity, and the condensation of the water vapour. The bees maintain a temperature of 20–30°C in the centre of a broodless winter cluster. This slowly warms the sphere to its outside, so that the outer bees do not become too cold. The bees slowly change position with one another in the cluster from inside to outside and from outside to inside.

An emergency

Sometimes fresh wax scales can be seen in the debris and occasionally also eggs that the queen possibly wanted to lay. If in winter the beekeeper finds round drone cappings — cappings of drone brood — he should assume that the colony has drone brood. This means that it no longer has a fertile queen. This can occur if she was not mated, or mated insufficiently, on the nuptial flight, or if this flight was prevented by a long period of cold weather. Possibly she has exhausted her sperm supply as a result of old age or the eggs are not being fertilized because of some internal abnormality. It could also be caused by the loss of the queen. Perhaps she is dead or the beekeeper accidentally injured her at the last inspection. In such an emergency situation the colony raises one or more pseudo-queens, or drone-laying workers. Workers are fed with royal jelly, which causes them to develop ovarioles of limited capacity and these start to produce eggs. But these eggs are not fertilized and thus only drones develop from them.

A colony that is just producing drone brood is doomed. It no longer has the ability to maintain the flow of life on its own. Only drones develop from the brood and the colony will collapse in the following spring at the latest. The beekeeper can take some remedial action by uniting it with another colony so that the remaining resources of the queenless one can be salvaged. But even this procedure is not always successful.

11. The Varroa Mite: A Parasite of Bee Colonies

Nowadays, beekeepers look for another symptom in the hive floor debris of their colonies. Our beekeeper soon points to tiny, dark-brown oval platelets between one and two millimetres in diameter that he has spotted on it. Thirty years ago, this phenomenon was unknown to beekeeping in Central Europe, but today there is hardly a colony that does not have these platelets and they can be found in hives over almost all of Europe. On their undersides are limb-like shapes that are not easy to see. Under magnification we can observe antennae, mouth-parts and four pairs of legs.

About 10–20 such platelets are present in the debris of our colony and the beekeeper explains that these are mites, arthropods called 'varroa mites.' *Varroa destructor,* formerly known as *Varroa jacobsoni* Oudemans, is a mite that was discovered and characterized on the East-Asian honey bee species *Apis cerana* at the beginning of the twentieth century. The mite lives in bee colonies. The female mites slip under the tergites, the rings of the abdominal segments, and suck out the haemolymph, the bee blood, of the worker bees.

Varroa mite (greatly enlarged) [From Herold and Weiss]

After a mite has matured for several days she leaves the bee and looks for a brood cell which is at two days before capping (7th to 8th day of brood development). She slips under the round larva into the liquid food and there enters a kind of paralysis. She ends this inactive state only when the food has been completely consumed and the larva in the capped cell has become a stretched larva. She feeds on the haemolymph of the larva for about two days. At intervals of thirty-hours minimum from about the fourth day after capping (13th day of brood development), a varroa female deposits on the cell wall so-called 'protonymphs' (developmental stage of the mite between egg and fully formed mite). These moult for the first time after a further thirty hours and move to the bee pupa in order to puncture it and suck haemolymph from it. The first protonymph is usually a male varroa. Still in the cell before the bee hatches, he then mates with subsequent fertile female mites (sisters) and then dies. The deposited protonymphs take about six days to develop into viable varroa.

In addition, about three varroa females can grow on a hatching worker bee, although usually only one of these mites is fertile. In drone brood, which of course requires 24 days to develop (14 days as capped brood), there may even be as many as between five and six varroa of which up to four may be fertile. After their proper period

Varroa on a bee pupa [From Herold and Weiss]

11. THE VARROA MITE: A PARASITE OF BEE COLONIES

Deformed bee [From Herold and Weiss]

of maturing on living bees, they move into the brood to reproduce again. Also, the varroa female can infest several brood cells during her life-span (6–7 are possible). It is interesting that even unmated varroa females enter brood cells to breed, but the protonymphs develop into males only. The female is then fertilized by the first sexually mature male. After that she can reproduce normally.

Bees hatching from infested cells are usually weakened. With higher varroa infestation two or more varroa may enter a brood cell. The hatching bees then show deformities of their wings or legs, or the brood even dies in the cell.

The reproduction of the varroa mite that parasitizes. *Apis mellifera* — the honey bee native to Central Europe and now distributed worldwide — generally takes a progressive course. This means that the number of mites in the colonies steadily increases. From a single infestation of a few mites, in the course of two or three years they become several thousand. Individual bees and brood of the infested colony are so weakened by this that other infections by bacteria or viruses lead to disease. Such colonies usually collapse in a short time in late summer. Apart from a few bees and dead brood the beekeeper suddenly finds empty hives.

The problem is especially serious because European honey bees have not yet been able to adapt their behaviour to limit the development of these mites. It is quite probable that honey bees would have been almost wiped out here had it not been for the fact that beekeepers, scientists and the chemical industry rushed to find ways of controlling mites in the colonies. It is now absolutely necessary to carry out varroa control measures on each hive at least once a year.

After the beekeeper has reinserted the sampling boards and made the entrances mouse-proof again, he makes a final survey of the apiary and its surroundings before departing. Further visits in January follow the same pattern.

12. Midwinter

At Christmas, beekeepers in former times would tell the bees of Christ's birth. Today this custom of communication between people and animals has been forgotten. But nowadays, even people who know about it and sense its deep significance often have difficulty in going about it in the right way. However, this encounter can be understood in the following way. We do not tell our bees and domestic animals about Christmas so that *they* may know Christ is born. We tell them each time so that *we* know it — that we know it *too*. This gives the animal world a feeling of confidence and security. And that is what being a domestic animal is all about, being trustingly commended to the care of human beings.

Beekeepers often put an ear to their beehives in winter to listen. Through the wood a soft humming, whispering, scratching and scraping can be heard. If the wood is knocked with the knuckles there is a roar from the colony and it soon goes quiet again. (Only a queenless colony moans its complaining reply with a longer rising and falling sound.) The colder it is outside the louder the humming. It is also louder when the colony has brood again. A winter cluster moves very slowly between the combs towards the stores thus gradually getting further and further away from the hive entrance. Now and then we notice a bee leaving the entrance and taking off, even in cold weather or when there is snow on the ground. She flies for a small distance and a few metres high only to drop to the ground or into the snow. She moves for a while and then quickly becomes torpid. If she falls on snow we notice that she sinks into it a little, sometimes one or two centimetres deep.

It is claimed that this phenomenon of bees flying out from time to time in winter occurs when they realize that they are going to die and leave so as not to burden the colony with their corpse.

Could this be the only satisfactory explanation?

13. Bee Christmas

The bees do not forage in winter time because they will not find anything for them in the frozen countryside. But the bee organism forms a broad sphere between the combs which beekeepers call the winter cluster. The bees pack themselves together and now and then take a tiny drop of their stores. The warmth of the summer sun is stored in the honey and it is set free in the body of the bees and so the bees are fed and warmed at the same time and guard their queen in their midst.

At the darkest time of the year, at midwinter, Christmastime, the bee organism dreams a dream of Christmas. It experiences the birth of Christ and through this dream renews its power each time to safeguard the secret of the light and warmth of the summer sun in the honey. Even more wonderful is this power in the beeswax. The beeswax candles burning on the Christmas tree ray out like small suns and spread with their light a magical scent.

Now and then the bees send one bee out of the winter cluster. She leaves the hive, walks across the alighting board and takes off into the bare, dull, cold winter landscape. She flies high into the air until she becomes torpid, falls to the ground and stiffens. With her life she blows a greeting of summer sun and bee humming from the bee colony into the wintry world: Spring is on the way!

(Winter contemplations; produced in a Main Lesson on beekeeping in Class 5 of the Jean-Paul Waldorf School, Kassel, Summer 1992)

IV. BUILD-UP

14. A Time of Awakening

On February 2, exactly six weeks after the winter solstice, the church celebrates Candlemas, the Feast of the Purification of the Virgin Mary. This is a significant date in farming and in phenomenological calendars for a number of reasons. At this time the sun is shining for an hour longer than it was at Christmas. It no longer gets dark so early in the afternoons. The songbirds have become noticeably more active and we can easily hear them singing again. Bud growth is visible on trees and shrubs and the stumps of trees and branches that have been cut around this time often begin to 'bleed' profusely. Plant sap is exuded and is frozen by the frost into little fountains on tree stumps.

After the brood-free period in the bee colony, the queen begins to lay the first eggs in the cells. She then soon starts to lay every freshly vacated storage cell increasing the area in an expanding circle so that the brood nest grows in the wake of the receding stores.

Cleansing flight

Sometimes the first mild days on which the sun heats the air around midday to over 12°C come in January, although they are usually at the end of February or the beginning of March. The hazel catkins often start to flower around this time and their yellow pollen is blown away by the wind.

When the beekeeper visits his apiary at this time, he removes the mouse guards and takes out the sampling boards to clean and inspect them. The first bees will probably have come out by now and soon many more will follow. There is an increasingly lively humming around the whole apiary, like a vigorous orientation flight of young bees. The bees fly in loops and lemniscatory forms of increasing size. In doing so they empty their recta, some in flight, some after alighting somewhere, especially on bright, gleaming surfaces. In residential areas, beekeepers have to ask neighbours who have hung their washing out to dry to take it in again. Otherwise they would wonder how the washing suddenly acquired a new pattern.

Bee foraging in spring [Photo: Herold and Weiss]

The bees even land on observers and deposit blobs that are ochre to brown in colour. The bees drag winter debris and the corpses of other bees out of the hive entrances. They fly off with them and drop them in the surroundings.

Meanwhile, the first forager bees have found their way to the hanging catkins on the hazel bushes and are gathering the bright-yellow pollen into their baskets. Soon the first returning foragers can be seen with yellow baskets at the hive entrances.

But as soon as it cools down, for example, when a cloud covers the sun, the flight traffic quickly stops and the bees hurriedly disappear back through the entrances. Otherwise the cleansing flight continues for at least another hour.

In the days that follow, bees can frequently be observed hurrying out of hive entrances. When it is mild they collect flower pollen and the first nectar from the early spring flowers, but even in bad weather they can be seen gathering water nearby. This is an indication of the increasing amount of brood being reared in the colonies. An inspection of hives at this time would show, in the centre of the bee cluster, a brood area that might extend to two or three frames.

The time of the cleansing flight is a joyful one for beekeepers. They see that their colonies have survived the winter so far and are at last becoming active in the open again.

15. Getting through Spring

After the cleansing flight, peace descends upon the apiary once more. It is still early enough in the year for the cold of night to set in again in the afternoon. The night often brings frost, especially when the day was sunny and clear. Perhaps it will even snow or hail the following day. The ground is still frozen where snow is lying.

Although the light is increasing day by day we cannot be sure of stable weather conditions, even after the spring equinox. After a brief warm period comes wind, rain, cold or snow. The lighting conditions change constantly — April weather.

The warmth of the sun can entice the bees out into the open and a short time later a cloud with a gust of wind and a shower of snow can throw them to the ground where they become torpid. If the sun comes out again and warms up the fallen bees they might stir again, but only a few of them become airborne once more and make it back to their hives.

Beekeepers watch these events with concern. They know well that getting through spring is a critical time that decides which colonies are healthy and strong enough to build up their resources through the Easter period and survive into summer.

Dangers

Inspecting the colonies at the beginning of March may present the following picture. The outside combs to the left and right are still full of capped stores after overwintering. Nearer the middle, the ring of remaining food has become smaller. There are patches of brood on two, rarely three, frames. If the weather in the preceding days was mild, and it was possible to bring in hazel and alder pollen, there would be a big patch of open brood. But if it was cold, even frosty, there would be less open brood. In this case the queen would have restricted her laying activity. If the weather stays cold for a long period and the bees therefore cannot go foraging, the queen stops laying altogether.

When it becomes so warm outside that the hive itself is warmed the bees begin to gnaw into the combs of stores at the outer margins of the hive, collect food and carry it to the inner region close to the sphere of brood. Depending on the consistency of the honey, more or less water is required for the bees to dissolve it and suck it up. Once the capping is removed the honey absorbs water from the cooler and more humid air around the margins of the hive, and thus becomes more pliable. If the honey stores have crystallized out to a greater degree and thus become harder it is more difficult for the bees to take them in again. That can depend on the forage brought in during the preceding year. Honey from early foraging (fruit-tree flowers, dandelion and especially oilseed rape) sometimes crystallizes very coarsely and becomes hard. But several summer nectar sources (sunflowers, mustard, *Phacelia*), especially a predominance of a single one, or certain sources from leaves (honeydew containing melezitose) can also become so hard in the combs that it is difficult, or even impossible, for the bees to liquefy the honey sufficiently in order to take it in as food. A colony affected in this way faces starvation because it cannot mobilize its stores. As a result it will only develop slowly.

There are other situations that are critical for the colony. When the total supply of stores is too small the bees will generally starve. The beekeeper can help them only to a limited extent by supplementary feeding, especially in cold weather. In any case the colony will come out of the winter weaker.

If the dome of stores over the bee colony is exhausted and the weather is extremely cold, the bees cannot move between one comb seam and another to get at the stores nearer the sides of the hive. In such situations they can starve even though they have full frames of stores nearby. The situation is especially critical when brood is already present and the increased scattering of the cluster among the combs draws it away from the brood because the bees must keep in contact with the food. This often splits the winter cluster apart. One part of the bee colony stays with the brood it is looking after and cools down and starves there, whereas the other — often significantly weakened — carries on feeding on the stores.

A beekeeper needs to find the cause of this development. Usually it is discovered that even before over-wintering there was not the right relationship between colony strength, comb development, hive volume and stores.

16. Winter Bees: Summer Bees

What developments are necessary in the beehive in spring? According to recent research (population estimates and counts) a bee colony starts the winter with an average of 12,000 bees (between 8,000 and 15,000). These bees hatch from the cells between August and October and form the winter colony. They survive the whole six months of winter until Easter. The last of the winter bees generally reach the end of their life-span at the beginning of May. This means that under certain circumstances winter bees can be eight months old whereas summer bees normally live for six weeks at most, about twenty-one days as house bees and a further 10–20 days as foragers.

If we study the development of individual summer and winter bees some marked differences can be noticed.

Summer bees live as house bees through more or less the full range of the necessary kinds of activity inside the colony such as heating, feeding, nursing, exuding wax and building, guarding the entrance, etc., and then, after a further period as foragers, die somewhere in the surroundings.

Winter bees hatch into a colony situation which can to some extent be described as a 'finished job.' The foraging activity is already very limited (less nectar available and the supply becoming gradually exhausted, shortening daylight hours, cooling outer conditions). The comb has already been built. The stores have been tended and sealed up. The brood needs to be fed less and less. Thus, the bees hatching into these circumstances do not have much to do externally. Although they still feed themselves up with 'bee bread' (pollen stored in cells mixed with other substances), they subsequently perform, to a limited extent, the tasks in the hive that the house bees would carry out in summer, in accordance with the development at that time. Instead, the winter bees form fat bodies, pads of stored protein and fat under the abdominal wall. More detailed research has shown that there are also changed hormonal conditions in the colonies at this time.

Only with brood rearing that starts at the end of winter (February–March) do individual bees begin a development comparable with summer bees. When this happens the stores in the fat bodies are broken down, nursing brood increases and building activities resume.

The more that new bees hatch out and go through their development as hive bees, the more winter bees become foragers and after a time they stay in the surroundings.

Population estimates show that a summer colony contains between 35,000 and 45,000 bees and rarely are there more than 50,000. Of those, at least one third are foragers, the rest are house bees. From towards the end of July (end of St John's Tide) the number of bees decreases and the brood volume is greatly reduced — approximately halved every four weeks until the last brood emerges in October [under UK conditions, small amounts of brood are not infrequently present in the winter months *Tr.*]. Then the summer bees gradually disappear. The winter cluster remains at the strength already mentioned and a healthy colony loses only a few bees until the time at which the increasing daylight stimulates the production of the first brood. Until that time, only a small portion of the stores is used. A brood free colony consumes less than a kilogram of stores per month. As soon as brood is being reared, fed and kept warm — it needs 35°C for healthy development even in a hard frost of long duration — the consumption of the stores increases to between four and six times what it was in brood-free conditions.

More water is used in colonies with brood, especially for dissolving the stores. Furthermore, there is increasing demand for fresh pollen. The early flowering trees are primarily hazel, alder and willow. The increasing foraging traffic combined with sudden changes in weather often leads to high losses of bees. The colonies fly themselves bare. If fewer young bees hatch in a colony than are lost, it becomes weaker. This has harmful repercussions for the situation inside the hive because the brood may be less well supplied and the queen reduces her laying. As a further consequence, even fewer young bees are produced three weeks later.

Help from the beekeeper

With constantly increasing development, the size of the brood nest doubles roughly every four weeks from mid-March to mid-May and the colonies then become very strong. Also the critical stage of loss of the winter bees is well matched by the hatching of young bees.

As the frames of comb can be moved at will, beekeepers have a number of options to help the colonies in their development. In

16. WINTER BEES: SUMMER BEES

colonies that are weak, the volume of the space can be reduced by removing empty unoccupied combs and separating the bees from the space vacated by inserting a movable dummy board (partition). The smaller space can be better managed as regards its warmth or climate by a weaker colony which, if it is healthy, is thus enabled to develop more quickly. Alternatively, in the case of a strong colony, the beekeeper may increase the volume of the space available to it by giving it empty frames to build comb in, or supplement the existing frames of comb with additional ones. He may also insert frames of foundation. These are thin sheets of beeswax ready embossed with the proper hexagonal pattern; that is, the same as those used to roll beeswax candles. These sheets are fixed vertically in the frames and the bees draw out the comb on them. The sheet of foundation forms the midrib of the comb.

Depending on the hive system in use there is also the option to increase the size of the hive above or below the brood nest. This can be done by adding empty frames or frames of foundation or comb in another box at the top or bottom. In the UK and USA this usually comprises one or more supers (boxes open at the top and bottom with rails to hold frames), which are placed on top of the brood box under the roof. A strong colony occupies this empty space, gradually fills it with fresh comb and uses it according to its capabilities and resources.

The beekeeper tries to get a picture of his colonies during each inspection. He observes what the hive is telling him: the flight traffic; the foraging; whether the incoming bees are also carrying pollen; the orientation flight of the young bees; the behaviour at the hive entrance, etc. Inspections in March are still infrequent, but in mild weather distinct differences can be observed between the colonies. The first drone flight is seen again around Easter, before the cherry trees are in flower.

17. Preparing for the Honey Harvest

At the beginning of the fruit trees blossoming, the beekeeper has a look at his colonies to see which ones will be mature enough for storing honey for harvest. When the brood space in which the bees over-wintered is well filled with bees, the brood nest has grown

Full comb of honey — all cells are capped
[Photo: Tim Kraus, Kassel 1996]

17. PREPARING FOR THE HONEY HARVEST

large and can obviously be supported satisfactorily, and the hive is really bursting at the seams, then it is high time to give the colony space to store honey. This is usually done by placing on top of the brood box another, shallower, box or 'super' with drawn comb, sometimes over a queen excluder. This is a sheet or grid of slots through which the queen cannot pass because of her larger body size. It prevents her from entering the super and producing brood there.

When the weather is favourable and lots of bees are visiting the early forage (flowers of fruit trees, dandelion, rape, sycamore etc.) this additional empty comb soon fills with honey. Sometime after Ascension Day or at Whitsun the beekeeper may harvest the first full honeycombs.

First the combs must be uncapped; that is, the wax cappings over the cells with ripe honey must be carefully removed with an uncapping fork or a knife. The honey can then be spun out in an extractor. The combs are placed upright in a basket which can rotate at speed in the extractor. The principle is similar to that of the spin-drier in a washing machine. Under the influence of the centrifugal force the honey flows out of the cells of the combs and flies into the extraction

Capped honeycomb is opened with an uncapping fork [Photo: Tim Kraus, Kassel 1996]

A four-frame tangential honey extractor [Photo: Tim Kraus, Kassel 1996]

vessel where it is collected. These spun-out combs can then be given back to the bees. If there is a good amount of summer forage, the beekeeper may be able to spin some more honey out, at the latest at the end of St. John's Tide (mid-July).

The honey harvest is always a festival for the beekeeper and his family.

18. Summer Inspections

Because of the inner life-rhythms of the honeybee, beekeepers always aim to inspect their colonies every nine days in the period from the beginning of May to the end of June. A beekeeper is familiar with the microclimate of the site of his apiary and keeps an eye on how the weather develops. He also has a certain amount of knowledge of the vegetation and its actual stage of development within flying range of his apiary.

When he first arrives at his apiary he tries to get a feel for the mood that prevails there. He studies the sky, wind, lighting and senses the quality of warmth. He sniffs the air and listens to the sounds. His gaze sweeps the trees and shrubs, taking in the status of the vegetation and looking for any departures from what he is accustomed to seeing. In this way he may also discover swarms hanging hidden from view. He observes the flight traffic, its direction, departures and arrivals, whether hunting about or direct, hesitating or hurrying, a little or a lot. If it is warm enough, then drones will be flying back and forth buzzing loudly from midday onwards. The orientation flight of the young bees shows that brood hatched in the colonies eight to ten days previously.

The ground in front of the hives is briefly carefully inspected too. A significant number of dead bees or deformed, small, wingless ejected brood, or bees incapable of flying, that is, crawlers and hoppers, tell the beekeeper about the health status of his colonies.

Has there been a poisoning because a farmer or gardener has misused insecticide on open flowers? If so, there will be piles of dead bees in front of the hives and little further activity in the poisoned colonies.

Some of the incoming bees are carrying fat, colourful pollen baskets. Yellow, orange and bright-green loads especially catch our eye. Several bees hardly manage to complete the return flight and they land first on the front of the hive above the entrance. After a pause they fly on again to disappear through the hive entrance or crawl head downwards down the wall into the entrance. Those bees returning without pollen are nectar gatherers or water carriers. A few bees are moving about on the alighting board facing outwards. They try to come into contact with incoming bees, briefly touching them

Beekeeping tools [From Herold and Weiss]

with their forelegs and antennae. Hardly a bee enters the hive without being checked.

At times there are bees to be seen at, and just inside, the entrance with their abdomens pointing outwards, bent and holding firmly onto the floor, fanning their wings powerfully. The easily felt stream of air that they drive out of the hive seems warm and humid and smells sweetly-aromatic, sometimes a little musty.

This behaviour tells the beekeeper that the bees have carried in a lot of nectar, which is being processed into honey by the bees in the hive.

Swarm control

A beekeeper knows the previous history and characteristics of his bees. Thus a few careful observations of flight behaviour and other activities at the entrances of some of the hives is sufficient for him to gain a good impression of how they are.

Before he opens the hives one by one, he perhaps makes a few notes of his impressions.

18. SUMMER INSPECTIONS

He sets out his equipment ready: a hive tool (a simple piece of metal with a chisel edge, and usually a hook for a multitude of uses such as scraping, shaving, levering, pushing and pulling); a goose feather or a flat hand-brush with a single row of soft bristles for sweeping bees off things when necessary.

Then the smoker is lit. Its fuel may be egg-tray cardboard or dried rotten wood or dried foliage, weeds (tansy) etc; in other words any plant material that will smoulder well and for a sufficient length of time and which, at a squeeze of the bellows, produces smoke on demand that should not smell unpleasant.

After the beekeeper has removed the roof and crown board (and insulation, if any) from the top of the hive there may be a plastic sheet or quilt underneath still to be removed. Before removing this the beekeeper may feel it to gauge the warmth coming from the bees below. It feels like body heat. The more powerful the colony, the further the warm patch extends from the middle to the edges.

Now. he gradually removes this last cover. A sticky, reddish-brown substance glues it to the frames. This is propolis, a sticky resin that the bees use to fill up all the cracks in the hive that they cannot crawl through themselves. All inner surfaces are thinly plastered with it and polished. The bees collect the resinous, waxy coating from leaves and flower buds — the bud resin — and process it into propolis by adding their own secretions and beeswax.

The beekeeper gives a few gentle puffs of smoke over the frames, which drives the bees downwards into the gaps between the frames. There can be seen the light-coloured surfaces of wax cappings, which shows that in recent days the bees have already produced some honey and stored it there. The beekeeper now pushes the frames together a little from the side in order to enlarge the gap at the side of the brood box. This allows him to lift out a comb, the second in from the outside, without squashing the bees that are closely packed together on it.

The comb has a beautiful ring of honey with white cappings at the top. Underneath is a large area of cells filled with a firmly compressed substance in a variety of colours, sometimes coated with a gleaming film. This is the pollen layer of the brood nest, a comb whose cells are filled with flower pollen collected from a variety of sources and covered with a film of honey to preserve it — 'bee bread.'

The beekeeper carefully puts the comb to one side and now removes the outermost comb which is almost completely filled with honey, a large part of it already capped. It is placed beside the first to be removed. On the next comb, the third in from the outside, there are cells of brood under a ring of honey and pollen, at this time filled with round larvae of various sizes. Therefore, the queen laid here between five and seven days ago.

Everywhere on the comb there are lots of bees; cleaning cells, feeding brood, processing honey or feeding each other. A few foragers are also visible, dancing to stimulate other bees to fly to the forage source that they have found.

The lower margin of the comb has the larger cell pattern of drone brood. There are fat round grubs in them too. The beekeeper replaces the comb in the empty space at the side of the brood box and lifts out the next comb. The first thing we notice here is the large, semicircular, capped patch of cells in the lower part of the frame — capped drone brood.

On top of that is a patch of light-brown, almost smoothly capped cells of worker brood, and under a ring of honey and pollen is a strip of as yet uncapped brood cells with fat round larvae in them. The keeper notices some small semicircular cones on the edge of the comb in a small hole by the wood of the frame. Their openings point downwards and still have a closed-in appearance. More such play cups — queen cups in the making — can be seen when the next frames are inspected. The last time the keeper looked at this hive he inserted two empty frames. They are already built in — a beautiful white comb moulded with worker cells during a 'flow' — and already have young larvae as well as eggs.

The beekeeper takes these findings as an expression of the vitality present in a colony that is keeping its options open. He notes that he will soon have to give this colony a super in which to store honey.

The beekeeper does not inspect every colony in such detail, that is, by checking every comb. In some of them he mainly inspects the frames he has added for the colony to expand. He only checks the colony more closely if he is not satisfied with what he finds there, for example, because it is no longer building comb or when the comb is not in a single piece but has been started in several places at once and therefore looks inharmonious.

Swarm mood

In one of the colonies that was already strong and vigorous early in the year, the super that was put on at the end of April is almost completely filled with fresh nectar and honey, part of which is already capped.

18. SUMMER INSPECTIONS

If the beekeeper tilts a comb, out of it fall a few drops of a bright, clear liquid that tastes sweetly aromatic. Fresh nectar is still rather runny.

The super is removed and carefully put to one side. A teeming mass of bees boils up from the gaps between the combs. Inspection reveals large areas of honey and pollen and a sizeable brood nest with brood at all stages: eggs, younger and older larvae and capped brood. There is also a lot of drone brood and lots of drones can be seen moving around on the margins of the combs. There are many cells with young bees just hatching from them.

It seems that every cell on every comb contains either stores or brood, and as a cell is vacated it is immediately cleaned and laid in again. Also, the frames recently given to allow expansion are now completely built out with comb. On the margin of the comb are some queen cups but this time they look opened out, and on the shiny, polished bottom of each is an egg. More queen cells that are already drawn-out to a greater length are discovered. In them is a whitish, shining substance with a small round larva in it. The greater the length to which the cell has been drawn out, the larger the larva. One of the cells is even longer and its opening has probably just been capped. A queen larva has a developmental period of three days as an egg and five days as a generously fed larva. A queen cell is capped eight days after egg-laying. Beekeepers know from this that on the next day a prime swarm will issue from the hive. The queen leaves the hive with a proportion of the bees and they swarm. The crops of the workers are filled with honey, otherwise the swarm will be taking nothing with it into an uncertain future. A week later the first young queen will hatch. Whether secondary swarms or casts will issue from the colony is determined by the bees, and depends on the circumstances inside and outside the hive at the time.

A vigorous colony in swarming mood may easily raise young queens in a couple of dozen swarm cells. When these hatch one by one several casts can issue from the hive in succession. Some of these may contain several young queens. If the beekeeper catches such a swarm and takes it into a cellar, when he later hives the swarm he may discover one or two dead queens on the floorboard. Obviously the swarm has killed them and thrown them out. Either that, or the queens have fought one another and the winner has survived.

Swarm control in beekeeping

Swarming is a time which the beekeeper can use to rejuvenate, select and breed from their colonies.

If he does not want to let the swarm issue, because it is always uncertain whether he will be able to find and take it afterwards, then he can prevent this by various means. First he needs to find the queen and remove her. He hides her away with a few bees in a small queen cage.

Then the bees from at least three quarters of the brood combs are shaken and brushed into a swarm-catching box fitted with a large funnel. This yields an artificial swarm weight of $1^1/_2$–2 kg. The queen is then released from the queen cage into it and the box may now be placed in a cellar.

Furthermore, an artificial prime swarm must be fed in the cellar within 24 hours at the latest as the bees will not have been able to take as much of their stores with them as a natural swarm. The feed is preferably candied honey, although sugar syrup, or, if necessary, fondant may be used. In addition, the composition of the bees is different, indeed arbitrary, according to which combs the beekeeper has shaken out into the artificial swarm. With a natural swarm, the bees that fly out of the colony have prepared for this by, amongst other things, gradually filling their crops. Thus a natural swarm can live for many days on the honey that it has taken with it.

A beekeeper can also prevent casts from issuing. Moreover, in this case, he does not wait for all the young queens to hatch naturally but carefully cuts the whole capped swarm cell out of the comb and lets them develop in queen cages. Then he creates artificial casts and gives one of the cells, or a hatched queen, to each.

These artificial swarms have to be fed. They too are hived in brood boxes with empty frames which the swarm will fill with comb.

However, he will usually restrict the space in the box with a dividing board or use a smaller box of three or more frames at first. The swarm can initially fully occupy this restricted space more readily and thus manage it better. This benefits the whole of its further development. The beekeeper then gradually increases the space with additional frames that are in turn filled with comb by the bees.

When the conditions are unfavourable outside it may be necessary to feed the swarms so that their development suffers no setbacks.

The young colonies develop in a similar way to that described in Parts II and III.

18. SUMMER INSPECTIONS

Moreover, the beekeeper will maintain a close watch on these swarms to check how they are getting on under his care, and how well they manage to balance the important relationships between colony strength, comb construction, quantity of brood and increasing stores. In doing so, he learns much that will be useful in later dealings with these colonies.

The pattern of reproduction and rejuvenation described above will particularly be carried out with colonies that, from their previous history over several generations of queens (queen lineages), the beekeeper regards as having good traits. For instance, he takes note of their health, vitality, strength, comb-building capacity, hygienic behaviour, eagerness to forage, honey yield, store development and management, docility, tolerance to being worked, and a number of other relevant characteristics, which may be important in the interaction between the beekeeper, the public and the bees.

Smoker [Photo: Tim Kraus, Kassel 1996]

V. A CHALLENGE

In a lecture on December 22, 1923, to the workers at the Goetheanum, Dornach, Switzerland, Rudolf Steiner described a particular aspect of the relationship between the human being and a swarm of bees:

> 'As I told you, when a young queen hatches in the hive there is something now amongst them which disturbs them. Previously the bees had lived in a kind of twilight. Now they see the young queen begin to shine. What is connected with this shining? It is connected with the fact that the young queen takes over from the old queen the power of the bee venom. And what the departing swarm fears, gentlemen, is that it will no longer have the bee venom. The bees are afraid that they will no longer be able to protect or save themselves. It departs just as the human soul does at death when it can no longer produce formic acid. The older bees depart when insufficient transformed formic acid, that is, bee venom, is present in the hive. And when we now look at the swarm, although it is of course visible, it looks like the human soul when

Swarm cluster [Photo: Rainer Vietor, Kassel 1995]

Swarm cluster from below [Photo: Rainer Vietor, Kassel, 1995]

it has to leave the body. An issuing swarm is a magnificent sight. Just as the human soul leaves the body, so also the old queen leaves the hive with her followers when the young queen is ready. And in the issuing swarm we can see a real picture of the departing human soul.

Gentlemen, this is truly wonderful! But the power of the human soul has never managed to go as far as making itself into little creatures. Nevertheless, there is always a tendency in this direction in us; always wanting to become like little creatures. We really have in us this constant desire inwardly to turn ourselves into crawling bacilli and bacteria, into tiny bees, yet we suppress it so that we can be fully human. But a hive of bees is not a complete human being. The swarming bees cannot find their way to the spiritual world. It is we who have to reincarnate them in a new beehive. This is a direct picture of human reincarnation. And anyone who has the opportunity to observe such a thing has an enormous respect for these swarming old bees with their queen who behaves as she does because she wants to go into the spiritual world. But she has become so physically material that she is unable to do this. And then the bees huddle together; become a single body. They want to be together. They want to be out of this world. As you know of course: whereas they otherwise would be flying, now they settle on the branch of a tree or on something else, snuggle up to one another, try to disappear because they want to go into the spiritual world. And then they become a proper colony once again when we help them by returning them to a new hive.'

(*Bees,* 8 lectures, Dornach, 3 Feb to 22 Dec 1923, GA 351, trans. T. Braatz. New York: Anthroposophic Press)

VI. THE GIFT FROM THE BEES

The Gift from the Bees
We work together all the day.
We work for others, not for pay;
So copy us if learn you will.
The sacred light from votive spill
Came from our crushed and melted comb
So think of this in your own home.
It should be said with great respect:
There is no work that we neglect.
The wax from off our abdomen
Perfumes the homes of bees and men.
We fashion cells in hanging tower,
Six-sided like the lily flower,
And store therein our flow'ry forage
That serves as bacon, eggs and porridge.
Their honey strengthens them and me:
So God, look kindly on the bee!

Traditional verse from heathland beekeeping
(recreated from the German by Chris Slade)

19. Honey

People, honey and bees

When most people think of honey, or of eating honey, they associate it with a feeling of security, abundance, comfort: the breakfast bread and honey on the morning of a holiday; warm milk and honey that calms the flitting thoughts after the stimuli of the daily hubbub, or soothing a hoarse throat and softening the pain.

The aura of honey is strong and our ideas of honey are deeply connected within us with a sense of a mysterious force at work.

Tasting honey — eating honey — embodies the joyful, amazing mood of confidence of a midsummer morning full of youthful freshness.

Eating a little honey each day strengthens our confidence for the day ahead.

Is it any wonder that *Homo statisticus germanicus* (the idealized average citizen of Germany) is the biggest consumer of honey in the world? In 2005 he ate between 1.1 and 1.5 kg on average. This is about 25 g per week, almost two level coffee spoonfuls.

Where does the wonderful power of honey come from? What is the basis of the aura that surrounds honey, which has survived like a tradition even to the present day?

Every child knows that when they see a bee on a flower it is collecting honey. But fewer and fewer people have really experienced the mystery of the little 'honey maker' (formerly *Apis mellifica,* now called *Apis mellifera* — 'honey bearer') or even looked into it in detail. But observing the processes carefully, enables us to learn something of the production of honey and ultimately of its significance for bees and people. It is not obvious that flowers contain a small amount of sweet fluid that is as clear as water. Only when the flowers are pulled apart do we really experience and comprehend what the bees do when they penetrate deeply into them. If we watch bees active in various flowers we will realize that their body shape is optimal for such activity. In order to discover that bee abdomens contain a crop in which nectar is collected and which is big enough to carry a significant amount, we first have to dissect them (see figure on page 56). And it takes many more observations to show that individual foraging bees are in no way directly self-motivated in their activity on the flowers.

A bee on her foraging flight usually visits flowers of the same kind (for example, apple blossom) in a small area, until her crop is full. Then she flies back to the hive. There she transfers the nectar to the house bees and soon flies off again. In doing so she flies to flowers of the same kind in the vicinity of the previous place as long as they are yielding nectar. Thus honey bees are constant in their preference for a particular flower species. Only when a source runs out — that is, when the secretion of nectar by the flowers visited so far decreases — are the bees then stimulated by the dances of other foragers in the hive to fly to a new, more attractive source of forage.

In the course of the year, the foraging zone of a particular colony changes. Whereas the flight radius remains rather short in spring, at the beginning of the availability of main nectar sources such as fruit blossom, dandelion, rape and, later, with the woodland forage or that of arable flowering crops (for example, *Phacelia*), it soon increases to 1,500 m, which corresponds to an area of 7 km^2, or at times even to 3,000 m, almost 30 km^2 or more. The flight radius is dependent on the strength of the colony, weather conditions and the attractiveness of a particular source of forage. The latter is predominantly dependent upon the flower density, the nectar yield of individual flowers, the sugar content of the nectar and whether this is easy to get at in the flower. The more favourable these factors, the more bees fly to particular sources of forage, even if they are further away.

The nectars of various flowers differ in sugar composition, aroma and other constituents. As a rule, foragers of a colony kept permanently in one location visit most of the species that flower in their flight range as long as they yield nectar and/or pollen. As a result there is constantly a mixture of nectars coming into the hive — more from one species, less from another — depending on what is available. The variety of nectars is dependent on the species diversity of flowering plants in the flight range.

As a rule, specific flower honeys (for example, fruit blossom, rape, chestnut, sunflower, etc.) can only be harvested by beekeepers who place their colonies in large expanses of a particular source of forage. Beekeepers move their bees to orchards, rape and sunflower fields, sweet chestnut blossom or other large honey-producing stands, as well as deciduous woodland and pine areas and, of course, heather moors soon after the flow begins at them. Migratory beekeeping also occurs; that is, beekeepers move with their bees.

The forager bees reorienting at an unfamiliar site will first find these plants and the bulk of the bees will subsequently visit only these.

Honey processing by bees

The nectar that has been brought in is generally only a slightly sweet fluid as clear as water, whereas honey has a viscous to solid consistency and is usually very sweet and aromatic. The smell and taste may be slightly sweet or very fruity, tart or even bitter-sweet. The colours of various honeys may range across the whole palette of earth colours from almost pure white, via light-yellow, deep-yellow, beige, ochre, golden-brown, brown, dark-brown to almost black. Nectar has a water content of up to 80% whereas honey is only 15–20% residual moisture. Nectar contains a variety of sugars of which several cannot be sources of energy either for bees or for people, because they cannot be digested in the form in which they occur. The sugars of honey (the term 'sugar' is here used in the same sense as in analytical chemistry) are primarily fructose and glucose plus a small amount of other types of sugar.

It may already be obvious from this information that the nectar brought into the hive must undergo special processing before it is changed into honey. (A comprehensive quantitative-analytical presentation of honey chemistry would take us beyond the framework of this discussion).

Once a forager has taken up sufficient nectar from the flowers, she commences the return journey to the hive. While she is collecting nectar she visits 100–200 or more individual flowers and enriches it with secretions from her royal jelly and salivary glands. These secretions are extremely rich in various enzymes, active substances which catalyse chemical reactions in living organisms. In this case they bring about, among other things, changes in the indigestible nectar sugars into the utilizable sugars, fructose and glucose. The enzymes also play a significant part in the beneficial effect of honey for people.

After their foraging flight, the foragers transfer the drops of nectar to the house bees. Subsequently, these drops are passed on many times from bee to bee and are further enriched by secretions from their glands. The more bees that participate in this processing chain the more concentrated the resulting honey. The nectar undergoes a significant change during the evaporation of its water content. It is actively concentrated by the bees. A bee engaged in honey processing ejects a droplet from its crop, spreads it out flat on its outstretched proboscis and then draws it together again. It repeats this procedure rhythmically every few seconds for up to 15–20 minutes. While this happens further glandular secretions are added. The honey that now soon becomes partially ripe is then spread in a thin layer in empty cells in the warmth of the brood nest area, or deposited in tiny droplets on the cell walls. In the dry warm air-stream (35°C, 40–50% relative humidity) that flows past the comb surface, more water is evaporated from the partially ripe honey until, after between one and three days, it falls to 20% residual moisture. The bees then carry it again to the margins of the comb, usually on the opposite side from the hive entrance. Once the cells are filled with ripe honey (less than 18% residual moisture) they are closed with an airtight wax capping. At the honey harvest the beekeeper can often find the same forage deposited in adjacent cells in patches of varying size. This shows the extent of the constancy of the bees' flower source.

In considering the overall phenomenon of honey production, two qualities can be experienced. The quality of the flowering plants in the environs of the hive are present in honey in a 'coagulated,' concentrated form. The substances are brought to a point in this way, not only spatially in the sense of flower diversity, but also temporally according to the sequence of flowering.

As a result, honey is a reflection of the relationship of each colony with its surroundings. Furthermore, it is an expression of the ability of the hive organism to manage the incoming nectar and to enrich it with its own secretions during the ripening activities.

Thus, the honey from a number of colonies neighbouring each other, and harvested by the beekeeper on the same day, but extracted, ripened and bottled separately, is likely to differ in appearance, aroma and consistency.

Sugar compounds	The average content of fructose is 38% and of glucose 31%. The content of maltose and other disaccharides and oligosaccharides is c. 9%
Vitamins	Honey contains traces of vitamins B1, B2, B6, C, biotin H, pantothenic acid, nicotinic acid and folic acid. It lacks the fat soluble vitamins.
Minerals	Potassium, magnesium, calcium, silicic acid, phosphorus, sulphur, manganese, silicon, sodium, copper, chlorine
Trace elements	Iron, copper, manganese, chromium, etc.
Amino acids	Glutamic acid, leucine/isoleucine, aspartic acid, phenylalanine, threonine, alanine, histidine, glycine, lysine, serine, valine, proline, cystine
Enzymes	Invertase, amylase, catalase, phosphatase, glucose oxidase
Hormones	Acetylcholine, growth hormones
Acids	malic acid, acetic acid, citric acid, lactic acid, butyric acid, succinic acid, pyroglutamic acid, gluconic acid, hydrochloric acid, phosphoric acid, formic acid.
Pollen	
Aromatic substances	Organic acids, phenylacetic acid ester, acetaldehyde, isobutyraldehyde, formaldehyde, acetone, diacetyl and a further 120 fragrance and aromatic compounds.

Table 3: Honey ingredients

19. HONEY

> Legal definition of honey (according to German honey regulations):
> 'Fluid, viscous fluid or crystalline foodstuff produced by bees by collecting flower nectar, other secretions of living parts of plants or secretions on living plants from insects, enriched and modified by the secretions of the bee's body, deposited in comb and ripened there.'

Returning forager bees transfer their loads to the young house bees

The young bees expose the nectar to the air by regurgitating it and taking it in again. This evaporates the water and as it passes the glands in the mouth substances important for the transformation of nectar into honey are added.

Stages in the process of regurgitation and taking in. The proboscis is slowly extended and swung back again.

1. Proboscis closed
2. A drop of nectar pumped back out of the crop appears in the glossal groove
3. The drop increases in size and partly flows downwards between the stipes and the galea
4. Further increase in the amount of nectar
5. Larger drop of nectar, held by the increasingly opened galea
6. Rhythmic opening and shutting of the galea in which the drop is repeatedly consumed and pumped out of the crop

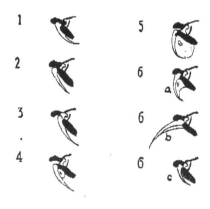

[From Büdel and Herold as well as Casaulta, Krieg and Spiess; originally from Park]

20. What is the Significance of Honey for Bees and People

Summer honey

It should first be made clear that a lot of honey is given as food in the raising of bee brood, whether directly in the feeding of the older brood, or in the form of royal jelly in the nursing of the young brood or the queen. However, with the royal jelly a completely new and transformed substance appears in the hive that no longer has anything in common with the original honey (see Chapter 21).

But the value of honey to bees can be further differentiated as may become clear in the following example.

If a bee has flown inside a house, and lies exhausted on the windowsill after long efforts to get out again, it will soon show no signs of life. But if someone notices this emergency situation and brings a drop of honey on their finger near the bee it will soon begin to lick it up. After a few seconds it will look more lively and begin to 'pump;' that is, visibly work its thoracic and wing musculature. Its wings will stretch out, tremble a little, and when sufficient honey has been consumed the bee will crawl away and finally fly off. This example may show that honey is the basis of all the outer activities of bees, for example, those activities that can be observed during foraging.

But the inside of the hive presents another picture: the bees can release the necessary heat for the healthy development of the brood at a temperature of 35°C by 'working' their thoracic and wing musculature while staying in one spot. This faculty of temperature regulation in the context of a colony is so strong that the necessary temperature of 35.5°C can be maintained precisely in the brood area when it is extremely cold outside, for example, in freezer room experiments at −35°C and even as low as −80°C, albeit with an increased consumption of honey. Likewise, the hive is ventilated by the buzzing, fanning of bee wings near the hive entrance and, if it is very hot, water is evaporated in the hive to cool it. Not least in requiring a plentiful supply of honey is the building of comb with wax exuded by the bees. Thus, in beeswax another completely new substance is created in the hive which has little in common with the substances used in its production.

If we can bring these facts together into an inner picture, we may realize that honey is an essential prerequisite not only for the outwardly visible and experiencable activities of the bees but also for the process of forming the colony as a basis for their physical life and manifestation of their nature.

'Winter honey'

Honey is also important to bees in its storage role. The period in which bees, in Central European latitudes (46°–57°N), can accumulate a surplus out of the surroundings of the hive from the end of April to the beginning of August comprises only a few whole days in many years. The rest of the time they are usually only able to bring in enough for their immediate use, and in bad weather they have to consume their stores, even in the summer. The beekeeper harvests part of the surplus and the remainder of it has to meet the requirement of the bees in the period of the year when they cannot bring in anything from outside. Bees do not hibernate although they overwinter with a certain amount of outward inactivity. But they do not do so as individual animals as do bumble bees, wasps or hornets, the queens of which enter into a winter paralysis in a hiding-place somewhere. Honey bees pass the winter months together as an organism in the hive. They are awake but only active within themselves. The reason why they store honey in significant amounts at all is that they need it to survive the winter. Their stores may be more than they need in which case we can harvest some of them.

If we examine the relative positions of brood nest and stores in a hive, we will notice that in all hive systems the nest is on the comb nearest to the hive entrance and the stores are further away from it.

What has been presented so far shows that honey is the result of the sense perception of the bee colony: sense perception in summer, smelling and tasting during visiting flowers in the environs — smelling, tasting and feeling during honey processing from gathered nectar in the hive. This sense perception is deposited in the form of honey in the combs. The bees consume it in winter and this reactivates the sense perception inside them. People refer to sense perception that is reactivated within as 'memory.' Likewise, we could say that honey has a 'memory' character, is memory. In winter it is for the bees a memory of their activity in the heat of summer — and an intimation, a foretaste, of the hot summer to come. This can be summarized in the following way: honey keeps the bees in the physical world.*

We have discussed the many manipulations that the nectar must undergo before it is stored as ripe honey by the bees in the cells of the comb. In this respect it should also be made clear that although the process of ripening nectar into the honey we consume takes place in the hive, it does not go through the metabolism of individual bees, nor become a new substance. The honey ripens within the hive, but outside the individual bees. It is concentrated and purified with bodily secretions of the worker bees and rhythmically moved and 'looked after' by them — potentized, so to speak.

The importance of honey to people

In his lectures on bees to the workers at the Goetheanum in Dornach, Rudolf Steiner said that bees are creatures that live primarily in smelling and tasting. If we study bees closely we can confirm this statement. The sense of sight is of somewhat secondary importance to bees. When a bee sets off foraging it does not search with its eyes. Stimulated by the dance of the other bees in the hive, the foragers know when they are flying off course and where they are heading (see Chapter 7). They do not have to search. They find their way to the flower site through an inner sense and once there follow their sense of smell. Then they taste whether they are on the right flower, and smell and taste as they gather nectar. The house bees smell and taste too as they process the nectar, add their secretions and concentrate the honey until they taste that it has become honey. The production of honey is the result of the smelling and tasting of the whole hive.

Rudolf Steiner indicated that consuming honey strengthens the human I-organization. This statement can be fully explained through studying the phenomenon of honey preparation in connection with the special sensory organization of bees. The human I-organization needs to assert itself constantly against the impressions of the senses.

---

* The author would like to draw the reader's attention to a delightful little book — *Frederick* by Leo Lionni, published by Alfred A. Knopf (USA) and Random House (Canada) — which illustrates this phenomenon of warming memories in a simple yet surprisingly profound manner.

Honey, a result of sense processes, is obviously a medium that helps the I-organization in asserting itself against the flood of sensorial stimuli, such that people do not function in the world according to their purely instinctive reactions.

The value of honey

When the comb is ripe it is harvested by the beekeeper. He extracts the honey mechanically from the combs by spinning or pressing. Afterwards he ripens the honey artificially by filtering, stirring and leaving it to stand until he finally bottles it to adorn our breakfast tables. 'Comb honey' is a delicacy. This is honey in its purest form, comparable with delicious confectionery, in its original packing of hexagonal cells, the comb freshly built by the bees that is brilliant white in colour and has not had brood in it.

Advice on the storage of honey: The processing of honey by the bees in the darkness of the hive occurs at a temperature of about 35°C. It is somewhat cooler on the margins of the comb, where ripe honey is stored in airtight capped cells. It is only warmed once more when the bees take it up into their organism again.

After the beekeeper has extracted the honey it cools down, is processed further by him and stored in jars which must be completely airtight. Then it needs to be kept in a cool, dark place, but not in a refrigerator, and then keeps its quality for an almost unlimited period of time. Light and warmth activate the enzymes in honey in the same way as should in fact happen when it is taken up into the organism. It is for that reason that honey has beneficial properties. If honey is kept in the light and heat, it will steadily lose its value. Also, when it is heated to make it more fluid, the enzymes are activated and it changes — the honey is altered and loses its effectiveness. Good beekeepers do not heat their honey after harvest and they put it in jars or in other containers for sale before its first setting.

A small calculation: On the foraging flight which lasts 30 to 45 minutes a bee visits 200–300 flowers of a plant species. In doing so it accumulates in its crop about 0.05 g nectar; that is, half its body weight. In really good weather a bee will manage about ten sorties a day which yields 0.5 g nectar. On a good day, 10,000 foraging bees on 100,000 foraging flights will visit 20–30 million flowers and in doing so may bring in about 5 kg nectar. This is processed into 1.5 kg honey in the hive.

What should honey really cost?
How much is a jar of honey worth?

Conversation with a queen bee

'If I politely, kindly, ask
For honey in a jar or flask

What would you charge me? How much please?
I can afford the best with ease.'

'You want a product pure and sweet?
You want the best that you can eat!

You're buying sunshine. Summer's through.
You know that honey's good for you!

There's nothing better to be found.
I'll calculate the price around

The work my bees did at the flowers.
It took them twenty thousand hours.

It's easy in the honey flow
And, yes, their pay is very low.

A pound an hour, or mark or yen;
That's twenty thousand of them then.

With twenty thousand pounds or more
I'm glad you're rich while we are poor!'

(Recreated by Chris Slade from the original German verse by Josef Guggenmos (1922) in *Was denkt die Maus am Donnerstag?* Munich: Deutscher Taschenbuch Verlag (dtv junior), 1980, 11th edition, p. 116f.)

One might justifiably add here that in his lectures on bees to the workers at the Goetheanum, Rudolf Steiner also said on the subject of honey: 'Honey is something so valuable that it is impossible to put a price on it.'

21. Royal Jelly: Liquid Feed for Queen Bees

Pollen and bee bread

Bees gather four different substances from the surroundings: sweetish, liquid nectar, floury pollen, sticky, bitter bud resin and water.

The process of gathering pollen by the bees is more visible than that of nectar and can be observed with the naked eye. Whereas pollen gatherers work more superficially on the flowers, nectar gatherers often have to force their way deeply inside the flowers in order to be able to reach the nectaries with their tongues, during which they are often also dusted with pollen. They frequently have to burrow, roll, dance right into the chalice of the flower as they manoeuvre, rip open and shake out the anthers with their mandibles and forelegs. The whole bee body is engaged in filling the pollen baskets and we soon notice the pollen remaining caught all over the hairs on the head, thorax and abdomen. At the same time, the free pollen is constantly wetted with fluid from the proboscis, for which purpose the bee has brought honey with her in her crop or uses nectar out of the flower. This sticks the pollen together somewhat.

As soon as the bee flies out of the flower, she begins to clean herself with her three pairs of legs and form the pollen baskets. The head is cleaned with the combs on the forelegs, the thorax with the middle pair. These also pick up the pollen collected by the forelegs. The forelegs are then scraped on the tarsal combs of the hind legs. Furthermore, these are rubbed together in parallel so that the pollen clumps are scraped out of the combs of one hind leg into the pollen rake of the other in a reciprocal process. The pollen baskets fill push by push. When this process is almost finished, it is packed down firmly from above and from the outside by the middle legs, during which fluid is again frequently pressed into the upper layer of pollen. The movements take place so quickly that we cannot easily follow them with the eye. But frequent observations over a short period of time enables a complete picture of the process to be built up (see also the illustrations for this on pages 14 and 15).

The pollen baskets vary in size and weight depending on the source and yield of the flowers. The average pollen load (two pollen baskets) weighs 8–12 mg and sometimes reaches 18–20 mg, depending on

Store comb; capped honey at the top, bee bread below it [Photo: Günther Friedmann, 1999]

the pollen available and its composition. The pollen colour differs between plant species and covers the entire colour spectrum. The bees show constancy of flower preference in pollen gathering too, and only rarely are the baskets multicoloured.

Returning bees scrape the pollen baskets into a comb cell by holding onto its upper margin, allowing their rear legs to hang deep into the cell, and then freeing the lumps of pollen from the basket surfaces. The pollen then falls to the floor of the cell. But here the constancy of flower preference stops. The pollen in the cells is generally of mixed origin, a mixture of colours.

The house bees who are responsible for this task further moisten the pollen mass and knead it. They 'stamp' it into layers in the cell with their heads until it is about three-quarters full. Then the mass is covered by a layer of honey and stays like that until the moment of use. In this condition the pollen undergoes a kind of mild lactic acid fermentation which preserves it to a limited extent and breaks it down for the digestion of the bees. The pollen stored in the comb in this way becomes what is called 'bee bread,' or also 'perga.'

Consumers can buy pollen baskets mixed in colours in jars, or in packets, either in shops, or directly from beekeepers. Beekeepers can harvest them by fitting a pollen trap, a plate with small round or star-shaped openings, to the hive entrance. The pollen gatherers must force themselves through these openings which causes them to lose some of their pollen baskets. The beekeeper cleans and dries the pollen, which would otherwise quickly spoil, so that he can offer it for sale. This flower pollen, in contrast to honey, has not yet undergone any development within the hive and is thus still to be regarded as raw material. Nutritional physiologists are quite unable to agree about whether human digestion can make significant use of this form of flower pollen. The individual pollen grains are covered with a tough outer skin, which is not easily broken down by the digestive juices.

It is different with the pollen in the comb, the bee bread. Here the fermentation process has already opened the pollen skin so that in fact all the valuable constituents of pollen can be digested. But this bee bread is rarely stored to excess in colonies, and it keeps only for a limited period of time there. Thus, it is most unusual that a beekeeper can harvest any of it, and it is a special blessing if we are ever personally presented with this bee gift by a beekeeper.

What is bee bread used for in colonies?

In Chapter 6 we described the typical development of a bee after it has hatched from a cell. We revealed that young bees frequently obtain food from the cells of stores. Close observation shows that in its first few days of life it really feeds itself up with honey and large amounts of bee bread.

At the same time it undertakes cleaning work in the hive, cleaning vacated cells and warming the brood by moving its wing musculature. Finally, it begins to feed the older larvae with a mixture of honey, bee bread and royal jelly. This shows that the royal jelly glands in the head of the young bee have started to release the secretion that one could also describe as bee milk. Appropriately, the bees tending the brood are referred to as nurse bees. After two or three days, the secretion of royal jelly reaches its peak and the nurse bees now feed just the young brood, which only receives royal jelly. Subsequently the secretion of the royal jelly glands diminishes and the wax glands develop in order to be able to exude the necessary quantity of wax scales to meet the construction requirements. By the time the period of wax work has ended, the venom gland has reached its maximum output and has filled the venom sac. As a rule, at this point the house bee period is over and the bee becomes a forager.

In scientific experiments, freshly hatched bees were isolated and put with a queen in an artificial colony. For food these received only honey or sugar solution and pollen from a pollen ageing experiment. The pollen spoiled very quickly and thus lost its nutritional value even for bees. As a result, the bees showed only limited activity and did not develop sufficient glandular secretions. The brood nests were very patchy because proper care of brood was obviously not possible. Also, the further familiar course of development of the individual bees could not take place properly.

From this somewhat crude negative evidence, we can understand that, for one thing, pollen provides the physical basis of nutrition for forming the bodies and structure within the colony. Also, the impression may arise that by consuming pollen the individual bees acquire the impulse to develop their intrinsic nature. That is essentially evidenced by the young bees, in connection with their consumption of pollen, performing the necessary jobs in the hive in a certain orderly sequence, and, in accordance with this, their glandular systems become active in a controlled sequence and the glandular activities cease again or become modified, for example, when the royal jelly glands become enzyme-secreting glands for processing honey.

Royal jelly

We return again in our studies to the subject of royal jelly. It is well known that this can be purchased for a not inconsiderable sum from health food shops etc. Royal jelly is a secretion of special glands in the bee head and is produced by young bees at a particular stage of their development, provided that they can consume sufficient honey and bee bread.

The name 'royal jelly' arises from the viscous, white, gelatinous plug in which the queen larva 'hangs.' The nurse bees feed a larva in a queen cell so generously with royal jelly that after the hatching of the 'princess' a lot remains dried up in the cell.

In the five days of her development before cell capping, the queen larva is fed about 1,600 times and constantly floats in a surfeit of food. The royal jelly that the workers and drones are fed with is clearer, less opaque, than queen royal jelly. Beekeepers only find these larvae floating in royal jelly in their first three days, but the worker and drone broods are given only as much food as they can consume. Thus a worker larva is fed only 143 times in the six days of her development, that is, approximately hourly. Apart from the bee brood, as the queen lays she is constantly offered royal jelly by the bees caring for her. In addition, she sometimes helps herself to honey from the store cells.

On many days between Easter and St John's Tide (June 24) in a healthy colony the queen bee lays an egg in a comb cell roughly every 40 to 60 seconds. Sometimes this amounts to 2,000 cells in a twenty-four hour period. The accumulated weight of these eggs greatly exceeds the body weight of the queen (0.16–0.23 g). She constantly produces new eggs in her body so long as she is fed sufficiently with royal jelly. If this flow diminishes or is interrupted, egg production diminishes in a short time or ceases altogether.

After three days small larvae hatch out of the eggs and are fed, warmed and tended by the nurse bees for a further six days (three days as young brood, three days as older brood). When the cell is capped a round larva of worker brood weighs about 0.15 g which is more than five hundred times the weight of the egg. In the case of 2,000 larvae of approximately the same age, this corresponds to a weight increase of 300 g in six days. This creative impulse constantly emanates from the queen through the bee colony by her egg-laying activity, and this is matched by the steady supply of royal jelly to the queen, which has been transformed and enriched from part of the substance collected in the surroundings.

Caring for the brood takes up almost half the period of development of the house bees; that is, from nine to ten of the twenty-one days. Although the productivity of nurse bees is distributed over several brood cells, on average a worker bee may raise the equivalent of two larvae with the royal jelly she produces. This shows the great metabolic output of the nurse bees. They create a completely new substance out of honey (which from the point of view of nutritional physiology works primarily as an energy supplier), and out of bee bread (which supplies protein, fat, starch, vitamins and mineral substances). This is the basis of the creative current that emanates from a laying queen and permeates the colony in rhythmic impulses (see page 52).

It is no wonder that people are so interested in royal jelly and invest it with energy-giving, health-promoting and life-prolonging properties. Indeed, a queen who is fed exclusively with royal jelly can sometimes live for as long as five years, and produce many hundreds of thousands of eggs. In contrast, in summer a worker that is fed for only three days of its larval stage with royal jelly and then a further three days with honey and bee bread may live for a maximum of only six weeks, and in winter perhaps reach an age of nine months. Thus, scientists are interested in finding out the as yet undiscovered constituents and properties of royal jelly.

But one thing is clear, the royal jelly is only to be found right in the middle of the bee colony, and where honey and bee bread are still recognizably similar to their floral origins, royal jelly is a pure creation of the bee organism.

Postscript

This book is intended to offer a foundation for understanding bee biology and the nature of the bee.

Describing bee biology on the basis of the observable phenomena is, for me, an exercise in understanding the nature of the bee. In doing so, I find that the most difficult challenge is to keep the presentation to what is in fact experienced. My method of phenomenological description, supplemented by information from professional beekeeping and scientific knowledge, is intended to help interested non-beekeepers (perhaps honey connoisseurs, teachers, doctors, etc) to get closer to the life of bees, and to make it comprehensible for them. But it is also intended to stimulate beekeepers and bee experts to look at bees again, perhaps in a different way. This study may stimulate a process that results in a deeper knowledge of bee nature. I should be grateful for any questions, advice, criticism, encouragement and help from readers who go through my studies with me.

I have increasingly found the many indications by Rudolf Steiner on understanding bees and the natural world to be of valuable help.

Michael Weiler
Deckenpfronn, 2006

Acknowledgments

I would like to thank the pupils at Jean-Paul School in Kassel. In 1992, I worked with them on a Main Lesson book on the theme 'Apiology,' which became the initial basis of this approach for me. I am also grateful to the participants of many introductory courses on the theme of 'Man and Bee' for their patience and their questions, which helped me to develop further phenomenological presentations.

I received many useful suggestions: from Dr Jochen Bockemühl; from the work of the Bee Study Group of the Science Section of the School of Spiritual Science, Dornach, Switzerland; as a result of my work with Illse Müller, Heidelberg, on 'Exercises and observations for approaching bee colonies;' and from the expert subcommittee on beekeeping at the Demeter Association.

Over a period of many years, Michael Olbrich-Majer of the Research Group for Biodynamic Agriculture in Darmstadt devoted many pages of the journal *Lebendige Erde* to my work.

Uli Nett, Kassel, for fifteen years a partner in our beekeeping venture, has been a real friend to me in my work.

Finally, and above all, I thank my wife, Monika, who in my years of beekeeping has patiently borne the stickiness of honey and my lack of punctuality, and who took care of our lively children while I looked after my bee colonies.

Tyll van der Voort of Oaklands Park Camphill Community had the initiative for the first courses on beekeeping in Britain. I want to thank him and Bernard Jarman of the Biodynamic Agricultural Association (BDAA) for their commitment and organization of the courses. Bernard was also the driving force behind having this book translated into English. David Heaf is a beekeeper, and this gave added depth to his translation. I would like to thank him specially for the excellent working relationship and communications during the production of this book.

Thanks to all at Floris Books for publishing the translation.

I would also like to thank the participants of the courses I was able to give in Britain. They enabled my wife and myself to get to know this wonderful country, and allowed me to discover that I can speak English for three days.

My greatest thanks must go to the bees: their images in me were obviously so strong that they overcame language barriers.

Selected Bibliography

Bruyn, Clive de, *Practical Beekeeping,* Crowood Press, 1997

Büdel, Anton & Herold, Edmund, *Bienen und Bienenzucht,* Munich, 1960

Casaulta, Glieci; Krieg, Josef; & Spiess, Walter (eds.), *Der Schweizerische Bienenvater,* Aarau, Switzerland, 1985

Davis, Celia, *The Honey Bee — Inside Out,* Bee Craft, 2004

Doering, Harald & Hornsmann, Erich, *Die Welt der Biene,* Munich, 1956

Herold, Edmund & Weiss, Karl, *Neue Imkerschule,* Munich, 1995

Hooper, Ted, *Guide to bees and honey,* Marston House Publishers, 1998

Park, W., *The storing and ripening of honey,* Rep. of the State Apiarists f. 1923 State of Iowa, 1924

Ruttner, Friedrich, *Naturgeschichte der Honigbienen,* Munich, 1992

Steiner, Rudolf (1923) *Über das Wesen der Bienen* (GA 351) Dornach (Switzerland), 1978

— *Bees,* 8 lectures, Dornach, Trans. T. Braatz. New York: Anthroposophic Press

— (1923), *Der Mensch als Zusammenklang des schaffenden, bildenden und gestaltenden Weltenwortes* (GA 230) Dornach (Switzerland), 1985

— *Harmony of the Creative Word* (GA 230) London, Rudolf Steiner Press, 2001

— (1924) *Geisteswissenschaftliche Grundlagen zum Gedeihen der Landwirtschaft* (GA 327) Dornach (Switzerland), 1979

— *Agriculture Course: The Birth of the Biodynamic Method* (GA 327), London, Rudolf Steiner Press, 2004

Winston, Mark L., *The Biology of the Honey Bee,* Harvard University Press, 1987

Zander, Enoch & Weiss, Karl, *Das Leben der Biene,* Stuttgart, 1964

Appendix: Demeter Beekeeping
by Günter Friedmann

A new approach to organic apiculture

In the summer of 1995 the Demeter Association introduced guidelines for the production of honey under Demeter certification. They were developed in collaboration with the Demeter Association's specialist subgroup on beekeeping.

The Demeter Association is the oldest organic farming organization in the world. In the UK, Demeter certification is managed by the Biodynamic Agricultural Association (www.biodynamic.org.uk). Its member producers, processors and traders are interested in doing more than just working close to nature without artificial fertilizers and plant protectants. Indeed, a special aim of this farming association is the care and development of soil fertility for which certain biodynamic preparations are used. Demeter principles and practice seek to put *culture* back into agri*culture*.

The Demeter guidelines for beekeeping have the same aims. The concern is not just the avoidance of residues in honey, wax and propolis, but keeping bees in accordance with their nature. This is the only way to ensure the productivity and vitality of bee colonies in the long term.

This aim is addressed explicitly in the foreword to the guidelines:

> 'beekeepers work in the context of biodynamics and orientate themselves primarily towards meeting the natural requirements of the colony. Management is so structured that the bee is able freely to unfold its true nature. Demeter beekeepers allow the colonies to build natural honeycomb. The basis for their reproduction, growth, rejuvenation and breeding is the process of swarming. Its own honey is the mainstay for supporting the colony through the winter.'

Demeter beekeeping in no way requires that bees forage only over land that is under organic management. The crucial thing about Demeter beekeeping is how the bees are managed.

It would not be surprising if many beekeepers find such statements unacceptable. They challenge fundamental assumptions of so-called modern beekeeping, such as systematic swarm prevention, making nucs (nuclei), artificial queen breeding and insemination. Many beekeepers believe that successful and profitable beekeeping is possible only with such measures. Yet an increasing number of beekeepers who work according to the Demeter guidelines, whether they have only a handful of colonies or several hundred, have found another way of dealing with bees that can also be successful and profitable.

The guidelines assert that like agriculture there is a *culture* which is particularly for beekeeping, that is, *apiculture*. However, it does not mean that beekeepers should do nothing and leave colonies to do as they please. On the contrary, they have a good many possibilities for intervention. But such interventions are not measured against maximizing honey yield but against the requirements of bee colonies themselves. Since healthy and populous colonies also bring in respectable amounts of honey, such an approach to beekeeping is also economically viable.

The Demeter guidelines focus on the way beekeepers manage their bees. Demeter beekeepers support a very consistent approach which avoids all systematic and arbitrary nuc production. Increase is only made via swarming which is the natural method of increase and reproduction. This does not mean that beekeepers have to stand around doing nothing and just wait until swarms are hanging in trees or flying away.

The important thing is not adhering rigidly to using all strong colonies for making up nucs and queen breeding, but intervening only in colonies which are intent on increase and reproduction because they are showing the symptoms of imminent swarming. They may then express their natural drive and go through the swarming process to a certain point. From the practical angle this means, first of all, that beekeepers must carry out regular swarming checks. If they then establish that some colonies are in a swarming mood they must not attempt to suppress this urge but make artificial swarms with the old queens of such colonies. The rest of the original colony may then be left or divided into nucs with swarm cells.

According to Demeter beekeepers, queens are not queens in the full sense of the word unless they have developed naturally; that is, they must be swarm or supersedure queens. Raising emergency queens is what bee colonies do when faced with the risk of dying out and should not form the basis of increase and reproduction. Dispensing with artificial queen breeding does not mean doing without selection altogether. Beekeepers are still free to choose only cells from good colonies. But the aim of selection should be to keep bees of European races that are adapted to the landscape and region and to avoid crossing them with bees from other continents.

Comb management is also part of the procedure. Demeter beekeepers regard comb construction as an integral part of a bee colony. The brood nest represents a natural unity. According to the guidelines, beekeepers should use only frames of a size that enables the brood nest to develop without hindrance from the frame bars (for example, Dadant size).

Apiculture that follows Demeter guidelines regards and treats the bee colony as a whole organism. Bees, brood, queen, stores, comb etc. all form a unity which should not be disrupted. Thus in the brood nest region, the central region of the colony, all the comb should be constructed by the bees alone — natural comb-building. The brood nest and the comb body, mediated by the natural comb construction, can then be appropriate for the size of the colony. Consequently, queen excluders are dispensed with.

Comb foundation may only be used in supers. This is a compromise between economy and ecology. Even so, the foundation used should be manufactured from natural comb. The reason for this requirement is connected with the quality of the wax. All kinds of residues accumulate in wax. Furthermore, wax deteriorates in the course of endless wax recycling. Naturally-made wax guarantees a not insignificant wax harvest and a wax of a high purity and quality. This finds an easy market as high quality wax is the best base for natural cosmetics.

Honey is the basis of the natural nutrition of bee colonies. Consequently, Demeter beekeepers need to add a certain amount of honey to winter feeds (at least 10% by weight). With this, and through the addition of salt and herbs, the nutritional quality is increased.

Honey from Demeter beekeeping should be bottled for sale before the first setting. This avoids the reduction in quality which happens at temperatures even below 40°C. The requirements of these guidelines pose significant challenges to beekeepers, especially at the time of the flower honey harvest, but they can be met by modern bottling technology. However, honey consumers also need to be helped to understand that during a particular part of the year only set honey is available.

The higher quality of Demeter honey warrants the higher price which, supported by the good reputation of the Demeter certification mark, is easily realizable.

All organic beekeeping organizations take for granted that *Varroa* mites should be controlled by formic, oxalic and lactic acids and not by artificial chemicals such as pyrethroids.

A beekeeper requiring Demeter certification must normally go through a conversion period in which they are allowed to sell honey as being from apiaries 'in conversion to Demeter' beekeeping. The entry criteria are such that beekeepers who already practise beekeeping to a particular organic standard may enter more easily, and may be able to develop this very different approach to beekeeping to full Demeter certification status within three years.

No doubt the expectations of the Demeter guidelines are high and demanding for beekeepers. Demeter recognition does not come on the cheap. The extra effort is worthwhile because implementation of the guidelines in one's own apiaries leads to a way of managing bees that is of an inherently high organic quality.

About the author:

Master Beekeeper, Günther Friedmann (b. 1956), is a professional beekeeper and manages his bees according to Demeter guidelines. He is a co-director of the Demeter Association in Germany and manages the organic production and processing standards inspections. His business is situated at Steinheim in the Swabian Alps and is a recognized training centre

Contact details

AUSTRALIA
Biodynamic Agricultural
Association
PO Box 54, Bellingen, NSW, 2454
Tel. +61 2 6655 0566
Fax.+61 2 6655 0565
poss@midcoast.com.au
www.biodynamics.net.au

CANADA
Demeter Canada
115 Des Myriques, CDN-Catevale
Q.C. J0B 1WO
Tel. +1 819 843 8488
laurier.chabot@sypatico.ca
www.demetercanada.com

GERMANY
Demeter-Bund
Brandschneise 1, 64295 Darmstadt
Tel. +49 61 55 84 690.
Fax +49 61 55 84 69 11,
Info@Demeter.de
www.demeter.de
www.demeter.net

IRELAND
Biodynamic Agricultural
Association
The Watergarden, IRL-
Thomastown, Co. Kilkenny
Tel/fax. +35 3565 4214
Email: bdaai@indigo.ie
Web: www.demeter.ie

NEW ZEALAND
Biodynamic Association
PO Box 39045,
Wellington Mail Centre
Tel. +64 458 953 66 Fax. 65
biodynamics@clear.net.nz
www.biodynamic.org.nz

SOUTH AFRICA
Biodynamic and Organic
Agricultural Association
PO Box 115, ZA-2056, Paulshof,
Gauteng
Tel/Fax. +27 118 0371 91
Email: Eleanor@pharma.co.za

UK
Biodynamic Agricultural
Association (BDAA)
The Secretary (BDAA), Painswick
Inn, Stroud
Tel/Fax. +44 1453 759501
bdaa@biodynamic.freeserve.co.uk
www.biodynamic.org.uk

USA
Biodynamic Farming and
Gardening Association
25844 Butler Road, Junction City,
OR 97448
Tel. +1 888 516 7797
Fax. +1 541 998 0106
biodynamic@aol.com
www.biodynamics.com

OTHER CONTACTS

Michaei Weiler
Imkerberatung@demeter.de
www.Bienen.de,
www.Imkerforum.de
www.beekeeping.com

Günter Friedmann
Bundesfachgruppe Demeter-
Bienenhaltung
Demeter-Imkerei, Küpfendorf 37
89555 Steinheim, Germany.
Tel. +49 7329 14 95
Demeter-Imker@demeter.de

For practical and authoratitive material (in English) on Varroa mite management:
www.apis.admin.ch

Book supplier specializing in bee books (new and second-hand):
Northern Bee Books, Scout Bottom Farm, Mytholmroyd, Hebden Bridge, HX7 5JS, UK.
Tel. +44 1422 882751
Fax. +44 1422 886157
sales@recordermail.demon.co.uk
www.beedata.com

Biodynamic Agriculture

Willy Schilthuis

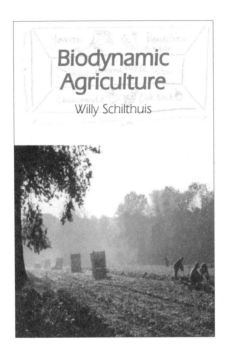

A concise and fully illustrated introduction to biodynamics. Biodynamics is an internationally recognized approach to organic agriculture in which the farmer or gardener respects and works with the spiritual dimension of the earth's environment.

In a world where conventional agricultural methods threaten the environment, biodynamic farms and gardens are designed to have a sound ecological balance. This book presents evidence that biodynamic crops put down deeper roots, show strong resistance to disease and have better keeping qualities than conventionally produced crops.

www.florisbooks.co.uk

Principles of Biodynamic Spray and Compost Preparations

Manfred Klett

A renowned biodynamic expert, Klett provides a fascinating overview of the history of agriculture, then goes on to the discuss the practicalities of spray and compost preparations and the philosophy behind them.

This is essential reading for any biodynamic gardener or farmer who wants to understand the background to core biodynamic techniques.

Based on keynote talks given by Manfred Klett at Biodynamic Agricultural Association conferences.

www.florisbooks.co.uk